# FUNDAMENTALS OF QUANTUM INFORMATION

# FUNDAMENTALS OF QUANTUM INFORMATION

**Hiroyuki Sagawa**
*University of Aizu, Japan*

**Nobuaki Yoshida**
*Kansai University, Japan*

World Scientific

NEW JERSEY • LONDON • SINGAPORE • BEIJING • SHANGHAI • HONG KONG • TAIPEI • CHENNAI

*Published by*

World Scientific Publishing Co. Pte. Ltd.
5 Toh Tuck Link, Singapore 596224
*USA office:* 27 Warren Street, Suite 401-402, Hackensack, NJ 07601
*UK office:* 57 Shelton Street, Covent Garden, London WC2H 9HE

**British Library Cataloguing-in-Publication Data**
A catalogue record for this book is available from the British Library.

**FUNDAMENTALS OF QUANTUM INFORMATION**



ISBN-13 978-981-4324-23-6
ISBN-10 981-4324-23-X

Printed in Singapore.

# Preface

Nobody objects to saying that the computer is one of the most important tools in our modern society, not only in scientific and industrial usage but also in everybody's daily life. According to Moore's law, microprocessors and memories of computers increase 4 times their speed and size every 3 years. With the development of technology, the extent of its participation in our life is far beyond what we imagined one decade ago. Its basic principles, however, remain unchanged for half a century. Information is stored, transmitted and processed by physics means. Thus, the basic principles of information and computation is based on the context of physics. Mathematics is its backbone. The question is then, after 20 years, how will computers be like? What are the ultimate physics limits to computations, communications and securities? The basic computer devices will be molecules or atoms governed by quantum mechanics where classical concepts such as the deterministic concept for physics phenomenon may not be applicable. On the contrary to the idea of classical computers, the basic principles of quantum mechanics begin to play an important role in recent progress of information theory. There have been rapid progress in communication technology, cryptography and data search algorithm based on the principle of quantum mechanics. For example, Shor's algorithm based on the idea of quantum computation will solve the factorization of large number within polynomial computation time. The communications will be transferred by the pairs of photons with unbreakable security. Thus, "quantum information" is noticed as the inevitable theory to describe the world of computers and communications in the 21th century.

The aim of this book is to provide fundamentals of main ideas and concepts of quantum computation and quantum information for both undergraduate and graduate students who have basic knowledge of mathematics and physics. The cross-disciplinary nature of this field has made it difficult for students to obtain the most important knowledge and techniques of the field. To overcome this difficulty, we try to describe all the materials in a self-contained way as much as possible. Firstly, we provide the basic ideas of quantum mechanics. Then it proceeds to describe main ideas of the quantum information theory and the quantum algorithm. Not sticking to mathematical strictness, we provide many concrete examples for clear understanding of the basic ideas and also for acquisition of practical ability to solve problems. We hope that this book will help students to open the door for a fascinating and unexpected computer world in the 21st century.

The structure of this book is in the following; after the introduction of historical summary of quantum mechanics and quantum information, the mathematical realization and the basic concepts of quantum mechanics are given in Chapters 2 and 3, respectively. A brief summary of "classical' information theory is given in Chapter 4. Chapters 5-10 are the highlight of this book devoted to the study of new ideas and concepts of quantum computations and quantum information. We try to give these materials in a self-contained manner as much as possible. Chapter 5 is devoted to quantum communication problems using an entangled state proposed by Einstein-Podolsky-Rosen. Chapters 6-9 will discuss quantum computation based on Shor's algorithm, quantum cryptography and quantum search algorithm, respectively. Physical devices of quantum computers are studied in Chapter 10. In the Appendix, the basic theorems of number theory are given.

We published a book titled "Quantum Information Theory", the first edition in 2003 and the second edition in 2009 (In Japanese, Springer-Japan, Tokyo). Some materials of this book are indebted to these Japanese books. We expand our Japanese book in various aspects and try to be more pedagogical and transparent. Especially the introductory materials of quantum mechanics are enlarged in Chapters 2 and 3 and "classical" information theory is also re-organized in Chapter 5. We include also recent progress in observations of EPR pair in Chapter 4. We put keywords at the beginning of each chapter and the summary is also given at the end of each chapter. The readers will be guided by these keywords and the level of understanding can be checked to read the summary of each chapter.

Problems are also given in each chapter. These problems will help the readers to obtain deeper understanding of the materials in this book. Solutions of all the problems are summarized at the end of this book.

June, 2010

Hiroyuki Sagawa
Nobuaki Yoshida

# Contents

# Chapter 1

# Introduction

The formulation of classical physics was started at the beginning of the 17th century by an Italian scientist Galileo Galilei. Through 17th century to 19th century, many experimental and theoretical findings were achieved by several brilliant scientists. The mechanics were established by Galileo Galilei and Issac Newton, the electromagnetism by Michael Faraday and James Maxwell and the thermodynamics by Ludwig Boltzmann. At the end of the 19th century, many physicists thought that all problems of the physical world were solved and the description of the universe was complete. However there were still few unsolved problems. Two of the major unsolved problems were Michelson-Morley measurement of speed of light and the black-body radiations. The study of two problems led to great discoveries in the history of science; special relativity and quantum mechanics. Albert Einstein contributed to both discoveries in 1905.

Classical physics can deal with the trajectory of particles such as falling bodies and motion of planets around the sun. The propagation of waves (light waves, sound waves) can be also treated in the context of classical physics. In classical physics, particles are assumed to have well defined positions and momenta. Conceptually, the position and the momentum of each particle are considered as objective quantities and can be observed simultaneously with the accuracy as much as you want. Classical physics has a deterministic character. A precise prediction of the evolution of the world is given by mathematical equations without any ambiguity. The existence of physical reality is assumed always independent of the observer.

Breakdowns of this conception was realized at the beginning of the 20th century. Lights behave not only as waves to be emitted and absorbed in bulbs, but also appear completely like particles. Max Planck introduced the quantization hypothesis (characterized by Planck's constant) for energy of light to obtain an unified equation to explain the black-body radiation.

Albert Einstein proposed in 1905 "Light Quanta" hypothesis to explain the photoelectric effect. On the contrary, electrons were considered as particles. It was found by Davisson and Germer that electrons display diffraction patterns as characteristics of waves. One of the most striking discoveries was that a hydrogen atom consists of a positively charged small heavy nucleus(proton), surrounded by a negatively charged light particle(electron). According to classical electromagnetism, the electron would emit radiations and collapse immediately into the nucleus.

In 1913, Niels Bohr proposed a model of hydrogen atom which explained both the stability of the atom and the existence of discrete energy spectra. His model was based on the quantum hypothesis of electron motion which conflicted with classical physics. In spite of its hypothesis, the Bohr model explained and predicted many atomic phenomena surprisingly well; the derivation of Rydberg constant and precise determination of X-ray energies. In 1925, Werner Heisenberg developed a quantum mechanical formalism in which the variables of position and momentum are presented by matrices and non-commutable. This idea was further developed to the uncertainty principle of two non-commutable observables in 1927. P. A. M. Dirac introduced the idea that physical quantities are represented by operators and physical states are vectors in a Hilbert space. In 1926, Erwin Schrödinger proposed a completely different idea for quantum mechanics: a differential formalism of quantum mechanics. Basic mathematical tools of the two formalisms look completely different, but the equivalence was proved mathematically.

Niels Bohr proposed the idea of statistical interpretation of quantum measurements to understand the duality of particle and wave. The statistical interpretation made a great success to understanding all the existing phenomena at that time without any ambiguity. Einstein, however, opposed strongly to Bohr's idea to claim "God does not play dice". He proposed many proposals of "Gedanken"(thought) experiments to deny the statistical interpretation. The famous one was the EPR (Einstein-Podolsky-Rosen) paradox, i.e., the problem of entanglement. Einstein proposed also a new idea of quantum mechanics called "Local Hidden Variable Theory". The debate between Einstein and Bohr appeared sometimes rather a problem of philosophy than that of physics since there was no proposal for measurable quantities to distinguish two theories. In 1966, J. S. Bell proposed the measurable quantities to distinguish two theories by Bell's inequality. This became a breakthrough of not only quantum measurement but also quantum teleportation. Einstein's theory was based on the idea of "action

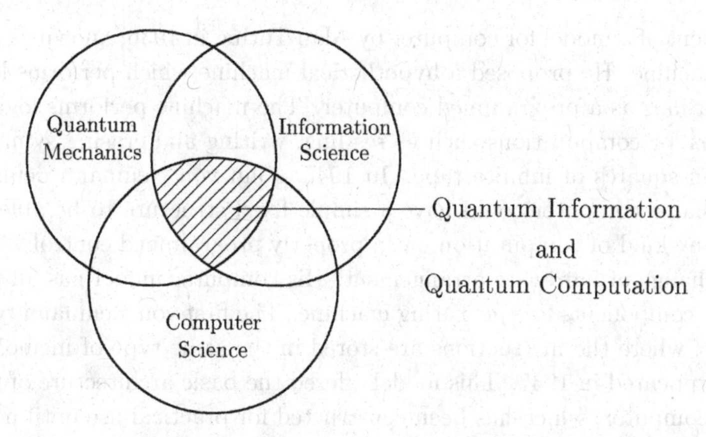


Fig. 1.1 Quantum information is a modern discipline of combining quantum mechanics, information theory and computer science and opens an window for promising future of computer science and technology.

Table 1.1 A brief chronological table of quantum physics and information science. Achievements of quantum physics are denoted by (QP), while those of computer science are labeled by (IS).

| Year | Name | Achievement |
|---|---|---|
| 1900 | M. Planck | Quantization of light energy (QP) |
| 1905 | A. Einstein | Light quanta hypothesis (QP) |
| 1913 | N. Bohr | Model of hydrogen atom (QP) |
| 1926 | E. Schrödinger | Wave equation of particle (QP) |
| 1927 | W. Heisenberg | Uncertainty principle (QP) |
| 1935 | A. Einstein, B. Podolsky, N. Rosen | EPR paradox (entanglement problem) (QP,CS) |
| 1936 | A. Turing | Turing machine (Automaton) (CS) |
| 1945 | J. von Neumann | Model of classical computer (CS) |
| 1966 | J. S. Bell | Bell's inequality (CS, QP) |
| 1984 | C. H. Bennett, G. Brassard | Quantum cryptography (CS) |
| 1985 | D. Deutsch | Model of quantum computation (CS) |
| 1994 | P. W. Shor | Quantum algorithm for finding prime factors (CS) |
| 1995 | L. K. Grover | Quantum search algorithm (CS) |

through medium", while Bohr claimed the idea of "action in distance" in quantum measurements. Eventually the measurement of Bell's inequality proved that the Bohr's idea was correct. Ironically, the denial of Einstein's local hidden variable theory opened the exciting new science field of quantum teleportation.

Another great intellectual success of the 20th century is computer science. The first milestone of progress on computer science should be the development of a model for computer by Alan Turing in 1936, known as the Turing machine. He proposed a hypothetical machine which performs logical operations as a programmed computer. The machine performs logical operations for computations such as reading, writing and erasing symbols written on squares of infinite tape. In 1945, John von Neumann demonstrated that a computer could have a simple fixed structure to be able to execute any kind of computation given properly programmed control without the change of hardware arrangement. His computer model has all the necessary components for the Turing machine. The first von Neumann type computer, where the instructions are stored in the same type of memories as data, appeared in 1947. This model offered the basic architecture of the classical computer, which has been constructed for practical use until now.

Computer hardware has grown in power at an extremely high speed ever since. Gordon Moore in 1965 claimed that computer power would double for constant cost every two years. This is known as Moore's law. Surprisingly, Moore's law has approximately held true in several decades since the 1960s. However, most computer scientists expect that this law will end soon in the early 21st century. The progress of computer technology is now facing fundamental difficulties of size. Quantum effects are beginning to interfere with electronic devices whose size becomes smaller and smaller. One possible solution of this difficulty is given by new computer paradigm based on the quantum mechanics called quantum computation. The idea is to use quantum mechanics to perform computations rather than the classical mechanics. In 1985, David Deutsch considered computer devices based upon the principles of quantum mechanics. He then proposed a simple example in which quantum computers might have much more computational powers than those of classical computers. Peter Shor demonstrated in 1994 the power of quantum computer in the problem of finding the prime factors of an integer. No efficient algorithm has been found so far to obtain the prime factors using the classical computers. The difficulty of factoring is the basic principle of the modern open-key cryptography called RSA cryptosystem. However, Shor's efficient algorithm for factoring on a quantum computer could be used to break RSA cryptosystem. This possible practical application of quantum computers to the break of cryptosystems has raised much of the interest in quantum computation and quantum information.

The power of quantum computers may not be limited to the mathematical problems such as the finding of prime factors of a large number. Further evidence for the power of quantum computers was given by Lov Grover in 1995. He showed that the problem of performing a search

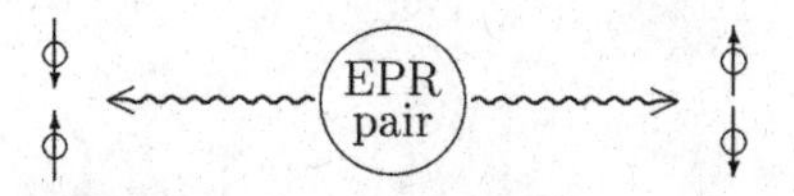


Fig. 1.2 Quantum mechanical entanglement. EPR pair is a typical example of entanglement of two particles.

through unstructured search space could also be speeded up on a quantum computer. In 1984, Charles Bennett and Gilles Brassard proposed a protocol using quantum mechanics to distribute keys between partners, without any possibility of wire tapping. This is called quantum cryptography and becomes the breakthrough of cryptosystems to open a window for unbreakable cryptography.

One of the most striking idea in quantum mechanics is the quantum entanglement. This problem was raised by the study of EPR pair measurement. In recent years there has been an enormous effort to understand the properties of entanglement. Many experimental studies were also done successfully using photons, electrons and nucleons. This intrinsic nature of quantum systems opens the door for possible future applications of quantum computations and quantum information.

Chapter 2

# Mathematical Realization of Quantum Mechanics

Any physical phenomenon is described by a state vector and an operator in the quantum world. We present mathematical methods which connect the microscopic physical world to the quantum mechanics. Two different representations will be discussed in this chapter. One is the matrix mechanics formulated by W. Heisenberg. Another one is Schrödinger wave mechanics. At first two models appear very different from each other, but it was proved that two theories are completely equivalent and lead to the same physical predictions. Both models have been quite useful not only in the formulation of quantum mechanics, but also in the development of quantum information theory.

**Keywords: state vector; Hermitian operator; Heisenberg's matrix mechanics; Schrödinger's wave function.**

## 2.1 State vector

A physical state in quantum mechanics is expressed by a vector $\psi$. It is also called a **state vector** or a **wave function**. The state vector belongs to a **Hilbert space** with any number of dimensions. It satisfies the standard algebra of vectors in Euclidean space such as addition, subtraction, multiplication and superposition. The state vector $\psi$ gives a completely new way of describing physical systems in contrast to the classical mechanics. It carries the information about all possible results of the measurements and replaces the classical concepts of position and momentum. In other words, quantum mechanics is not a deterministic theory, being different from Newtonian mechanics. Quantum phenomena will be predicted only by probabilities.

The **superposition** of two state vectors $\psi$ and $\varphi$ makes another state vector,

$$\Psi = c_1\psi + c_2\varphi \tag{2.1}$$

in the same Hilbert space. In Eq. (2.1), $c_1$ and $c_2$ are complex numbers. Notice in Euclidean space, coefficients for the superposition of vectors are always real numbers. In quantum mechanics, it will become inherently complex numbers as will be seen later (see the section of Schrödinger equation). The idea of superposition of two states is a fundamental and unique concept of quantum mechanics different from any properties of classical mechanics.

Suppose a superposed state $\Psi$ in Eq. (2.1) being a combination of spin up and spin down state with coefficients $c_1$ and $c_2$, respectively. A single measurement of the system will be either spin up or spin down state. The quantum theory never predict the spin direction of single event. Alternatively, performing a sufficient large number of measurements $N$, a number of events to observe spin up states $N_\uparrow$ and that of spin down state $N_\downarrow$ are predicted to be

$$\frac{N_\uparrow}{N} = |c_1|^2 \quad \text{and} \quad \frac{N_\downarrow}{N} = |c_2|^2, \tag{2.2}$$

by the coefficients $c_1$ and $c_2$. The principle of quantum mechanics makes possible a superposition of two photons moving opposite directions: a combination of spin up and down states (a classical analogue of this state is a

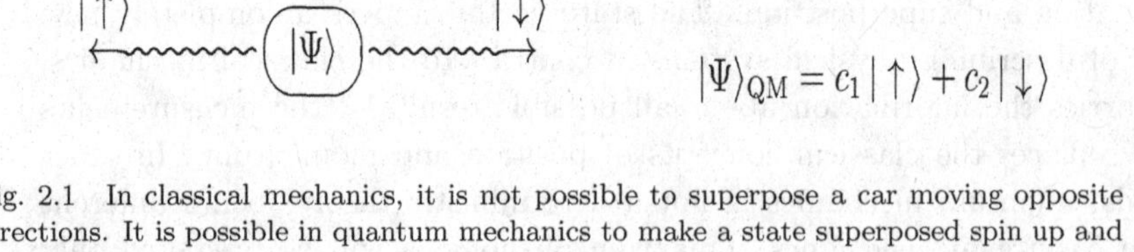

Fig. 2.1 In classical mechanics, it is not possible to superpose a car moving opposite directions. It is possible in quantum mechanics to make a state superposed spin up and spin down states.

combination of clockwise and anti-clockwise spinning top). These properties were never found in the classical world, for example, a linear combination of a car moving opposite directions: a spinning top rotates clockwise or anti-clockwise, but not as a superposition of both. One can imagine that one observes a car moving to the right with 50% probability and moving to the left with 50% probability?

The wave function $\psi$ is often expressed as

$$|\psi\rangle, \tag{2.3}$$

which is named a **ket vector**. This notation was invented by P. A. M. Dirac. The **Hermitian conjugate** vector to a ket vector is called the **bra vector** and expressed as

$$\langle\psi| = (|\psi\rangle)^\dagger \tag{2.4}$$

The bra and ket vectors satisfy the normalization condition:

$$\langle\psi|\psi\rangle = 1. \tag{2.5}$$

Any state vector $|\psi\rangle$ can be expressed as a linear combination of basis vectors $\{|i\rangle; i = 1, \ldots, n\}$ in a Hilbert space,

$$|\psi\rangle = \sum_i |i\rangle\langle i|\psi\rangle = \sum_i a_i|i\rangle; \quad a_i \equiv \langle i|\psi\rangle, \tag{2.6}$$

where the **orthonormal condition**:

$$\langle i|j\rangle = \delta_{i,j} \tag{2.7}$$

is required between any two elements from the complete set $\{|i\rangle\}$. The state vector may be expressed by a matrix form with the amplitudes $a_i$ filling the successive rows of a column vector,

$$|\psi\rangle = \begin{pmatrix} a_1 \\ a_2 \\ \vdots \\ a_n \end{pmatrix}, \tag{2.8}$$

with the understanding that $n$ can be infinity ($n \to \infty$) in the Hilbert space. A bra vector is written in a matrix form by a row vector

$$\langle\psi| = (a_1^*, a_2^*, \ldots, a_n^*). \tag{2.9}$$

The normalization condition of a state vector $\psi$ is written as

$$\langle\psi|\psi\rangle = \sum_i a_i^* a_i = \sum_i |a_i|^2 = 1. \tag{2.10}$$

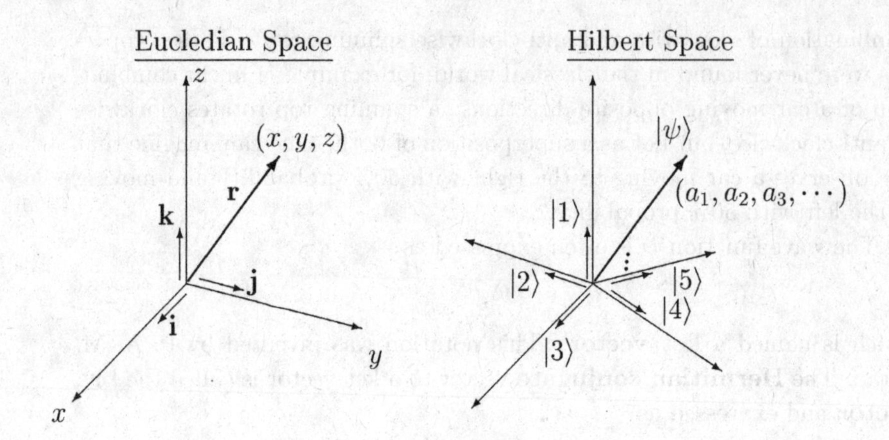


Fig. 2.2 A position vector in Euclidean space and a state vector $|\psi_i\rangle$ in a Hilbert space. The $|i\rangle$ is the basis vector in a Hilbert space. The dimension of Euclidean space is finite (three for the position vector), while that in the Hilbert space will be infinity: $\{|i\rangle; i = 1 \cdots n\}$ .

The matrix form of quantum mechanics is called "**Heisenberg's matrix mechanics**" in contrast to "**Schrödinger's realization**" by a wave function. The matrix mechanics is quite useful to deal with the spin degree of freedom of elementary particles.

A useful set of basis states $|i\rangle$ is given by the column vectors with all their amplitudes being zero except the value 1 at the i-th row. With these base vectors, the state vector $|\psi\rangle$ is expressed as

$$|\psi\rangle = a_1 \begin{pmatrix} 1 \\ 0 \\ 0 \\ \vdots \\ 0 \end{pmatrix} + a_2 \begin{pmatrix} 0 \\ 1 \\ 0 \\ \vdots \\ 0 \end{pmatrix} + a_3 \begin{pmatrix} 0 \\ 0 \\ 1 \\ \vdots \\ 0 \end{pmatrix} + \cdots + a_n \begin{pmatrix} 0 \\ 0 \\ 0 \\ \vdots \\ 1 \end{pmatrix}. \tag{2.11}$$

The **unit matrix I** in the Hilbert space is expressed as

$$\mathbf{I} = \sum_i |i\rangle\langle i|, \tag{2.12}$$

which is an $n \times n$ matrix generated by a tensor product of a row vector and a column vector.

## 2.2 Operator

Any physical quantities, or observables, that can be measured by experiments, are expressed as functions of the coordinate $\mathbf{r}$ and the momentum $\mathbf{p}$ in classical mechanics; any physical observable A, B, $\cdots$ will be written as functions of coordinate $\mathbf{r}$ and momentum $\mathbf{p}$. In classical mechanics, the physical quantities are $c-$numbers and commutable,i.e., AB=BA. It is not the case anymore in quantum mechanics. In quantum mechanics, observables $A, B, \cdots$ (denoted by italic symbols hereafter) are written by operators: the numbers (the coordinate and the momentum) are replaced by operators, A$\rightarrow A(\mathbf{r}, \mathbf{p})$, which act on the state vector. This procedure makes possible one to one correspondence between classical and quantum observables. There are also purely quantum mechanical observables such as the spin operators.

When an operator $A$ acts on a state vector $\psi$ belonging to a Hilbert space, it transfers to a new vector in the same Hilbert space,

$$\varphi = A\psi. \tag{2.13}$$

When two operators $A$ and $B$ act successively to the state $\psi$, either $BA\varphi$ or $AB\psi$ these operators obey a non-commutable algebra, $BA\psi \neq AB\psi$ as for the case of rotation in the three dimensional Euclidean space. If the vector $A\psi$ is proportional to $\psi$, the vector $\psi$ is called an **eigenvector** of the operator $A$. The proportional constant $a_i$ is called the **eigenvalue**;

$$A\psi = a_i\psi. \tag{2.14}$$

The scalar product of a vector $A\psi_1$ and another vector $\psi_2$ is called the matrix element

$$\langle\psi_2|A|\psi_1\rangle. \tag{2.15}$$

In a matrix representation, an operator is expressed as

$$A = \begin{pmatrix} \langle 1|A|1\rangle, & \langle 1|A|2\rangle, & \ldots, & \langle 1|A|n\rangle \\ \langle 2|A|1\rangle, & \langle 2|A|2\rangle, & \ldots, & \langle 2|A|n\rangle \\ \langle 3|A|1\rangle, & \langle 3|A|2\rangle, & \ldots, & \langle 3|A|n\rangle \\ \vdots & \vdots & \vdots & \vdots \\ \langle n|A|1\rangle, & \langle n|A|2\rangle, & \ldots, & \langle n|A|n\rangle \end{pmatrix}, \tag{2.16}$$

for any set of the basis vectors $|i\rangle$ with $\{i = 1, n\}$.

The **expectation value**, or the **average**, of an operator $A$ for a state vector $|\psi\rangle$ is given by

$$\langle A\rangle \equiv \langle\psi|A|\psi\rangle \tag{2.17}$$

Table 2.1 Notations of quantum mechanical state and operator. The notation of a bra and a ket vector was invented by P. A. M. Dirac.

| Notation | Description |
|---|---|
| $c^*$ | Complex conjugate of a complex number $c$ |
| (example) | $c^* = (a+bi)^* = a - ib$ ($a$ and $b$ are real numbers) |
| $\|\psi\rangle$ | State vector in a Hilbert space. a ket vector |
| $\langle\psi\| = (\|\psi\rangle)^\dagger$ | State vector in a Hilbert space. a bra vector |
| $\langle\varphi\|\psi\rangle$ | Scalar product of state vectors $\|\psi\rangle$ and $\|\varphi\rangle$ |
| $\|\Psi\rangle = c_1\|\psi\rangle + c_2\|\varphi\rangle$ | Superposition of $\|\psi\rangle$ and $\|\varphi\rangle$ |
| $\|\psi\rangle \otimes \|\varphi\rangle$ | Tensor product of $\|\psi\rangle$ and $\|\varphi\rangle$ |
| (example) | $\begin{pmatrix}1\\0\end{pmatrix} \otimes \begin{pmatrix}1\\0\end{pmatrix} = \begin{pmatrix}1\\0\\0\\0\end{pmatrix}, \quad \begin{pmatrix}1\\0\end{pmatrix} \otimes \begin{pmatrix}0\\1\end{pmatrix} = \begin{pmatrix}0\\1\\0\\0\end{pmatrix}$ |
| $A$ | An operator |
| $A^\dagger$ | Hermitian conjugate of an operator $A$ |
| (example) | $A^\dagger = \begin{pmatrix} \langle 1\|A\|1\rangle, & \langle 1\|A\|2\rangle, & \ldots, & \langle 1\|A\|n\rangle \\ \langle 2\|A\|1\rangle, & \langle 2\|A\|2\rangle, & \ldots, & \langle 2\|A\|n\rangle \\ \langle 3\|A\|1\rangle, & \langle 3\|A\|2\rangle, & \ldots, & \langle 3\|A\|n\rangle \\ \vdots & \vdots & \vdots & \vdots \\ \langle n\|A\|1\rangle, & \langle n\|A\|2\rangle, & \ldots, & \langle n\|A\|n\rangle \end{pmatrix}^\dagger$ |
| | $= \begin{pmatrix} \langle 1\|A\|1\rangle^*, & \langle 2\|A\|1\rangle^*, & \ldots, & \langle n\|A\|1\rangle^* \\ \langle 1\|A\|2\rangle^*, & \langle 2\|A\|2\rangle^*, & \ldots, & \langle n\|A\|2\rangle^* \\ \langle 1\|A\|3\rangle^*, & \langle 2\|A\|3\rangle^*, & \ldots, & \langle n\|A\|3\rangle^* \\ \vdots & \vdots & \vdots & \vdots \\ \langle 1\|A\|n\rangle^*, & \langle 2\|A\|n\rangle^*, & \ldots, & \langle n\|A\|n\rangle^* \end{pmatrix}$ |
| $\langle\varphi\|A\|\psi\rangle$ | Scalar product of $\|\varphi\rangle$ and $A\|\psi\rangle$ |
| $\langle\varphi\|A^\dagger\|\psi\rangle = \langle\psi\|A\|\varphi\rangle^*$ | Condition for Hermitian conjugate operator |

as a diagonal matrix element of Eq. (2.15). Any measured value of the physical observable must be a real number in classical mechanics. In quantum mechanics, this condition is undertaken equivalently as the expectation value $\langle\psi|A|\psi\rangle$ being a real number. The **Hermitian conjugate** $A^\dagger$ is defined as

$$\langle\psi_2|A^\dagger|\psi_1\rangle = \langle\psi_1|A|\psi_2\rangle^*. \tag{2.18}$$

An operator $A$ is called **Hermitian operator** if it satisfies the condition $A^\dagger = A$. For the Hermitian operator, the matrix element becomes

$$\langle\psi_2|A|\psi_1\rangle = \langle\psi_1|A|\psi_2\rangle^* \tag{2.19}$$

for any pair of state vectors, $|\psi_1\rangle$ and $|\psi_2\rangle$. Alternately, one could say that Eq. (2.19) requests the operator $A$ being Hermitian. The expectation value (2.17) is a real number for any state vector if Eq. (2.19) is satisfied.

## 2.3 Time evolution of wave function

We now implement the state vector by a complex function of the coordinate and time $\psi(\mathbf{r}, t)$, in which the spatial dimension appears explicitly. This is called **Schrödinger's representation** of wave function in contrast to Heisenberg's matrix mechanics. Two realizations of quantum mechanics at first looked very different and made a lot of confusions in the beginning when two theory proposed in 1920th, but eventually it was proved that two theories are identical in quantum mechanical predictions. It is just a matter of convenience which theory will be adopted. The state vector in the Schrödinger representation is called a **wave function.** The time evolution of the wave function $\psi$ is determined by the time-dependent **Schrödinger equation,**

$$i\hbar \frac{\partial \psi(\mathbf{r}, t)}{\partial t} = H\psi(\mathbf{r}, t), \tag{2.20}$$

where the Hamiltonian $H$ is the sum of the kinetic energy and the potential energy, while the symbol $\hbar = h/2\pi$ stands for the **Planck constant** $h$ ($h = 6.63 \times 10^{-34}\mathrm{J \cdot s}$) divided by $2\pi$.

We will introduce an operator $U(t', t)$, which will evolve the wave function $\psi(t)$ to $\psi(t')$,

$$\psi(t') = U(t', t)\psi(t). \tag{2.21}$$

For an infinitesimal time $\Delta t$, we denote the evolution operator as

$$U(t' = t + \Delta t, t) = U(\Delta t). \tag{2.22}$$

By applying the definition of the differentiation,

$$\frac{d\psi(t)}{dt} = \lim_{\Delta t \to 0} \frac{\psi(t + \Delta t) - \psi(t)}{\Delta t}, \tag{2.23}$$

Eq. (2.20) can be rewritten for infinitesimal time $\Delta t$ as

$$\psi(t + \Delta t) = \left(1 - \frac{iH\Delta t}{\hbar}\right)\psi(t) \tag{2.24}$$

in the case of time-independent Hamiltonian. To compare with a relation $U(\Delta t)\psi(t) = \psi(t + \Delta t)$, we obtain

$$U(\Delta t) = \left(1 - \frac{iH\Delta t}{\hbar}\right). \tag{2.25}$$

from Eq. (2.24). Since the Hamiltonian is Hermitian, $H^\dagger = H$, we have the relation

$$U(\Delta t)U^\dagger(\Delta t) = \left(1 - \frac{iH\Delta t}{\hbar}\right)\left(1 + \frac{iH\Delta t}{\hbar}\right) \cong 1 \tag{2.26}$$

valid up to the first order in $\Delta t$. Equation (2.26) shows that $U(\Delta t)$ is a unitary operator $U^\dagger = U^{-1}$. For a successive time evolution, $t_0 \to t = t_0 + \Delta t \to t' = t + \Delta t$, the wave function $\psi$ can be written as

$$\begin{aligned} \psi(t') &= U(t', t)\psi(t) = U(t', t)U(t, t_0)\psi(t_0) \\ &\equiv U(t', t_0)\psi(t_0). \end{aligned} \tag{2.27}$$

Then $U(t', t_0)$ is expressed as a product of infinitesimal time-evolution operators,

$$U(t', t_0) = U(t', t)U(t, t_0). \tag{2.28}$$

Furthermore, an equation

$$U(t', t_0) = U(\Delta t)U(t, t_0) = \left(1 - \frac{iH}{\hbar}\Delta t\right) U(t, t_0) \tag{2.29}$$

gives

$$\frac{U(t', t_0) - U(t, t_0)}{\Delta t} = -\frac{iH}{\hbar}U(t, t_0), \tag{2.30}$$

which is rewritten to be

$$\frac{1}{U(t, t_0)}\frac{dU(t, t_0)}{dt} = -\frac{iH}{\hbar}, \tag{2.31}$$

in the limit of $\Delta t \to 0$. By integrating both sides of Eq. (2.31) from $t_0$ to $t$, we obtain

$$U(t, t_0) = e^{-\frac{i}{\hbar}H(t-t_0)}, \tag{2.32}$$

where the initial condition $U(t_0, t_0) = 1$ has been taken into account. The expression (2.32) will give a time evolution from $t_1$ to $t_2$ as

$$\psi(\mathbf{r}, t_2) = U(t_2, t_1)\psi(\mathbf{r}, t_1). \tag{2.33}$$

## 2.4 Eigenvalue and eigenstate of Hamiltonian

For a time-independent Hamiltonian, let us assume that $\psi$ can be expressed as a product of a time-independent function $\phi(\mathbf{r})$ and a time-dependent function $f(t)$:

$$\psi(\mathbf{r}, t) = \phi(\mathbf{r})f(t). \tag{2.34}$$

Substituting (2.34) into (2.20), we obtain

$$i\hbar\phi(\mathbf{r})\frac{df(t)}{dt} = H\phi(\mathbf{r})f(t). \tag{2.35}$$

By dividing (2.35) by $\psi(\mathbf{r}, t)$, we have

$$i\hbar \frac{1}{f(t)} \frac{df(t)}{dt} = \frac{1}{\phi(\mathbf{r})} H\phi(\mathbf{r}) \ \ (\equiv E), \tag{2.36}$$

in which the time-dependent part and the position-dependent part are separated. The equation (2.36) is satisfied only if both sides are constant. Writing the constant as $E$, we can integrate $f(t)$ as

$$f(t) = e^{-i\frac{E}{\hbar}t}. \tag{2.37}$$

The equation for the position-dependent part, on the other hand, becomes

$$H\phi(\mathbf{r}) = E\phi(\mathbf{r}), \tag{2.38}$$

indicating that $E$ is an **eigenvalue of the Hamiltonian**, and $\phi(\mathbf{r})$ the **eigenstate**. Thus, the solution of the Schrödinger equation (2.20) can be separated into the position-dependent and the time-dependent parts:

$$\psi(\mathbf{r}, t) = \phi(\mathbf{r}) e^{\frac{-iEt}{\hbar}}. \tag{2.39}$$

In general, the energy $E$ becomes **discrete** if $\phi(\mathbf{r})$ is confined in a finite region, while it takes continuous values if $\phi(\mathbf{r})$ extends to infinity. The eigenstate of the time-independent Hamiltonian is called the **stable state**. Since the energy is the eigenvalue of Hamiltonian, there exists a lower limit. The state with the lowest energy is called the **ground state**, while other states are called **excited states**. The latter states are further distinguished as the first excited state, the second excited state, $\cdots$ according to the order of energy.

If the atom or the molecule is isolated, and if there is no external perturbation, the system stays at the lowest-energy state, i.e., in the ground state. If any external energy is given into the system, the ground state can be changed into some excited state. Then, it will come back to the ground state again after emitting the energy. The transition between energy levels, however, occurs only if the law of energy conservation is satisfied. In other words, the change of the states can happen only if the incoming energy is equal to the difference between the two energy levels, and the outgoing energy must be equal to the energy difference between the energy levels so that the emission satisfies the energy conservation. These energy selection rules play essential roles in the implementation of **quantum bits**.

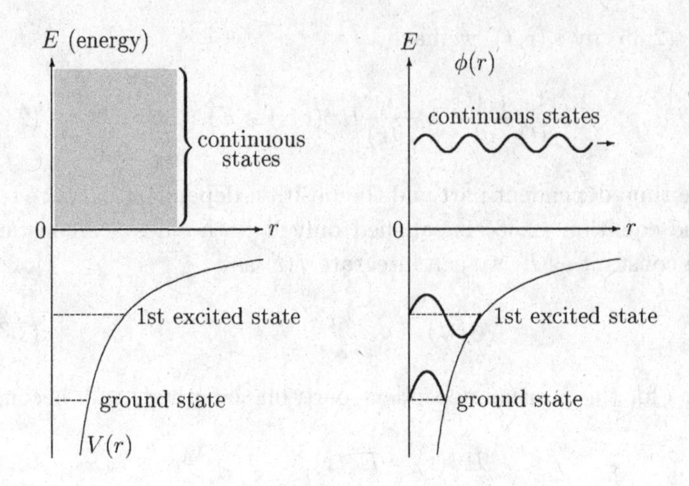


Fig. 2.3 Energy levels and wave functions.

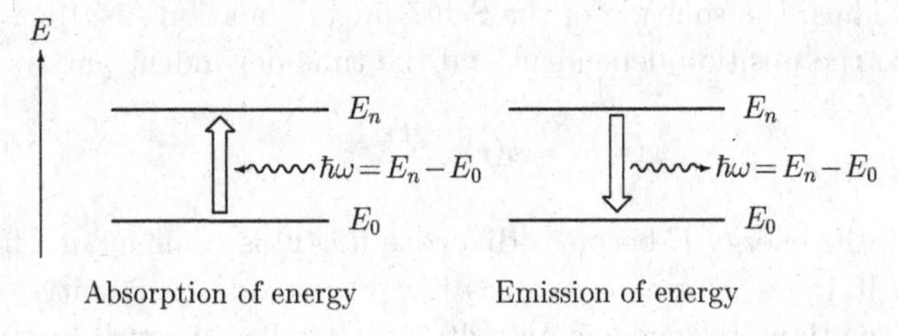


Fig. 2.4 Energy selection rule in absorption and emission of energy between two energy levels.

## Summary of Chapter 2

(1) The state of physical system is described by a state vector $\psi$ in a Hilbert space.
(2) Every physical quantity is expressed by a Hermitian operator.
(3) A state vector and an operator are described by matrix forms in the Heisenberg's realization of quantum mechanics.
(4) The time evolution of the system is governed by a time-dependent Schrödinger equation

$$i\hbar\frac{\partial\psi}{\partial t} = H\psi,$$

where $\psi$ is called a wave function.
(5) The eigenvalue of Hamiltonian is discrete when the wave function is confined in a finite region.

## 2.5 Problems

[1] Find which combinations of the ket vectors are linearly independent among

$$|\psi_1\rangle = \begin{pmatrix} 1 \\ 0 \end{pmatrix}, \ |\psi_2\rangle = \begin{pmatrix} 0 \\ 1 \end{pmatrix}, \ |\psi_3\rangle = \begin{pmatrix} i \\ 1 \end{pmatrix}, \ |\psi_4\rangle = \begin{pmatrix} -i \\ 1 \end{pmatrix}.$$

a) $|\psi_1\rangle$ and $|\psi_2\rangle$?
b) $|\psi_1\rangle$ and $|\psi_3\rangle$?
c) $|\psi_2\rangle$ and $|\psi_4\rangle$?
d) $|\psi_3\rangle$ and $|\psi_3\rangle$?

[2] Find eigenstates of operators $\frac{d}{dx}$ and $\frac{d^2}{dx^2}$ in the following functions;

$$\text{a) } e^{ax}, \ \text{b) } e^{ax^2}, \ \text{c) } ax+b, \ \text{d) } \sin x.$$

[3] Show that a superposed state $\Psi = a_1\psi_1 + a_2\psi_2$ is also a solution of the Schrödinger equation, if $\psi_1$ and $\psi_2$ are two solutions of the Schrödinger equation

$$i\hbar\frac{\partial}{\partial t}\psi = H\psi.$$

The coefficients $a_1$ and $a_2$ are arbitrary complex numbers.

[4] Consider two eigenstates of $\psi_1$ and $\psi_2$ of momentum operator $p$ with eigenvalues of $k$ and $-k$

$$p\psi_1 = k\psi_1, \ p\psi_2 = -k\psi_2,$$

and the system is prepared in the superposed state,

$$\Psi = \frac{1}{\sqrt{2}}(\psi_1 + \psi_2).$$

a) Calculate the expectation value of momentum for $\Psi$.
b) Calculate the expectation value of kinetic energy $p^2/2m$.

# Chapter 3

# Basic Concepts of Quantum Mechanics

It is known that a particle shows two different characters, particle-wave duality in the quantum mechanical world. Photons show the character of wave, such as diffraction and interference, but the photo-electric effect and the black-body radiation can be explained only by the particle character of light. In the microscopic world, the duality is the property not only inherent in photons but also common in all particles including electrons, protons and neutrons. From this peculiar character of particles, the motion in the microscopic world can be understood only in the probability interpretation, i.e., the particles are described by the wave function $\psi$ with a wave character and the measurements give the probabilities of finding particles in space-time. This is the concept called the Copenhagen interpretation of quantum mechanics, developed by Niels Bohr and his group. In this chapter, we summarize the basic ideas, principles and concepts of quantum mechanics.

**Keywords: probability interpretation; uncertainty principle; spin; quantum bit (qubit).**

## 3.1 Wave function and probability interpretation

The **wave function** to describe the particle motion in three-dimensional space is expressed as a function $\psi(\mathbf{r}, t)$ of the coordinate vector $\mathbf{r} = (x, y, z)$ and time $t$ in a Schrödinger representation. The quantum states of a particle may also have discrete quantum numbers such as spin and angular momentum. The wave function $\psi$ is also called the **probability amplitude**, and its square:

$$|\psi(\mathbf{r}, t)|^2 = \psi^*(\mathbf{r}, t)\psi(\mathbf{r}, t) \tag{3.1}$$

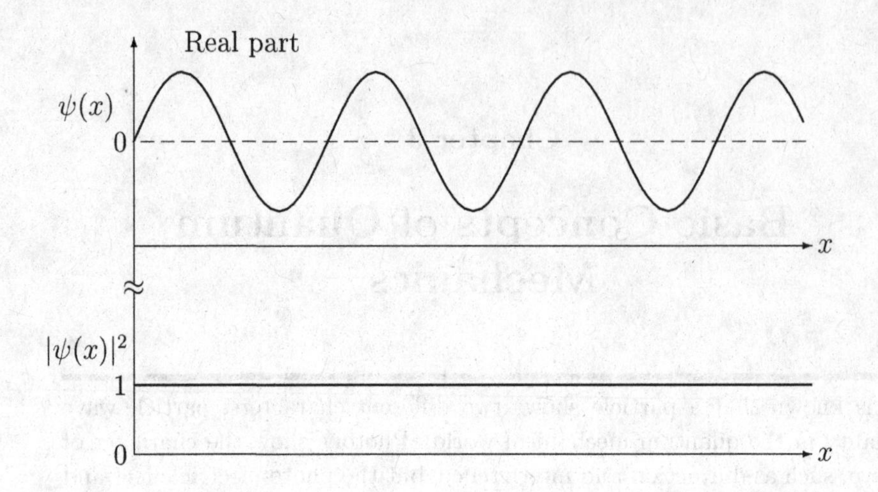


Fig. 3.1 A plane wave $\psi(x)$ and its probability $|\psi(x)|^2$.

gives the probability density of finding the particle at position $\mathbf{r}$ at time $t$. The wave function is normalized so that its integral over the whole space is unity, In. Eq. (3.1), $\psi^*$ stands for the complex conjugate of $\psi$.

## 3.2 Commutation relation and uncertainty relation

In quantum mechanics, the position $\mathbf{r}$ and the momentum $\mathbf{p}$ act as operators upon the wave function $\psi(\mathbf{r}, t)$. The momentum operator is expressed by the differentiation $\nabla$ as

$$\mathbf{p} = -i\hbar\nabla.$$

For any pair of operators $A, B$, the **commutation relation** is defined as

$$[A, B] = AB - BA. \tag{3.2}$$

For $\mathbf{r}$ and $\mathbf{p}$, the following relations hold:

$$[x, p_x] = i\hbar, \quad [y, p_y] = i\hbar, \quad [z, p_z] = i\hbar. \tag{3.3}$$

The relation between the operators given by Eq. (3.3) is **non-commutable.** Other commutation relations with $\mathbf{r}$ and $\mathbf{p}$, such as $[x, y]$, $[p_x, p_y]$, $[y, p_x]$, are all 0, and are called **commutable.**

---

**Example 3.1** Show $[x, p_x] = i\hbar$.

**Solution** The operator $p_x$ is written as $p_x = \frac{\hbar}{i}\frac{\partial}{\partial x}$. By applying the commutation relation $[x, p_x]$ to any wave function $\psi$, we have

$$\begin{aligned}[x, p_x]\psi &= (xp_x - p_x x)\psi \\ &= \frac{\hbar}{i}\left(x\frac{\partial}{\partial x}\psi - \frac{\partial}{\partial x}(x\psi)\right) \\ &= \frac{\hbar}{i}\left(x\frac{\partial}{\partial x}\psi - \psi - x\frac{\partial}{\partial x}\psi\right) \\ &= i\hbar\psi.\end{aligned}$$

Thus $[x, p_x] = i\hbar$ has been derived.

---

We consider two operators that are independent of each other. Let us show that the operators $A$ and $B$ are commutable if all the eigenstates of $A$ are at the same time the eigenstates of $B$. Taking a common eigenstate $\psi$, whose eigenvalue of $A$ is $a$ while that of $B$ is $b$:

$$A\psi = a\psi, \qquad B\psi = b\psi. \tag{3.4}$$

The commutation relation $[A, B]$ with $\psi$ gives

$$[A, B]\psi = (AB - BA)\psi = (ab - ba)\psi = 0. \tag{3.5}$$

Since the common eigenstates of $A$ and $B$ constitute the complete set, the equality $[A, B] = 0$ holds. Conversely, it can be shown that there exist the complete set of common eigenstates of $A$ and $B$ if $[A, B] = 0$.

How about the case where $A$ and $B$ do not commute? Let us consider, in particular, the case with $A = x$ and $B = p_x$. If there were a common eigenstate $\psi$ of $A$ and $B$, it would satisfy the equation: $[A, B]\psi = 0$. This equation contradicts the relation obtained from the commutation relation: $[A, B]\psi = i\hbar\psi$. Thus it is concluded that there is no common eigenstate of the position $x$ and the momentum $p_x$. In other words, no state can have a definite position and a definite momentum simultaneously. This conclusion implies the fact that if $x$ is determined by measurements without any ambiguity, large fluctuation (uncertainty) arises in the measurements of $p_x$. Alternately, if $p_x$ is determined precisely, $x$ fluctuates seriously and its values can not be determined. This relation is called the **uncertainty relation** between position and momentum, and is one of the most fundamental principles of quantum mechanics proposed by Werner Heisenberg.

Let us define the uncertainty in the observation of an operator $A$ as the standard deviation $\Delta A$,

$$(\Delta A)^2 \equiv \langle (A - \langle A \rangle)^2 \rangle = \langle A^2 \rangle - \langle A \rangle^2. \tag{3.6}$$

The uncertain relation is between position and momentum is expressed as

$$\Delta x \cdot \Delta p_x \geq \frac{\hbar}{2}, \qquad \Delta y \cdot \Delta p_y \geq \frac{\hbar}{2}, \qquad \Delta z \cdot \Delta p_z \geq \frac{\hbar}{2}. \tag{3.7}$$

---

**Example 3.2** Assume that two Hermitian operators $A$ and $B$ satisfy the commutation relation $[A, B] = i\hbar$. Prove the uncertainty relation $\Delta A \cdot \Delta B \geq \frac{\hbar}{2}$

**Solution** Define the following state vector by applying two Hermitian operators $A - \langle A \rangle$ and $B - \langle B \rangle$ to the state vector $|\psi\rangle$:

$$|\tilde{\psi}\rangle = (A - \langle A \rangle)|\psi\rangle + i\lambda(B - \langle B \rangle)|\psi\rangle, \tag{3.8}$$

where $\lambda$ is a real number. The bra vector $\langle \tilde{\psi}|$ conjugate to the state (3.8) is written to be

$$\langle \tilde{\psi}| = (|\tilde{\psi}\rangle)^\dagger = \langle \psi|(A - \langle A \rangle) - i\lambda\langle \psi|(B - \langle B \rangle). \tag{3.9}$$

The scalar product of the bra (3.9) and the ket (3.8) becomes

$$\langle \tilde{\psi}|\tilde{\psi}\rangle = \langle \psi|(A - \langle A \rangle)^2 + i\lambda[A, B] + \lambda^2(B - \langle B \rangle)^2|\psi\rangle. \tag{3.10}$$

Since the scalar product (norm) of the same state vector must be nonnegative, (3.10) satisfies the condition

$$\langle \tilde{\psi}|\tilde{\psi}\rangle = (\Delta A)^2 - \lambda\hbar + \lambda^2(\Delta B)^2 \geq 0. \tag{3.11}$$

Note that the inequality hold with any real number $\lambda$. Considering (3.11) as a quadratic equation with respect to $\lambda$, its discriminant must be negative or zero:

$$\hbar^2 - 4(\Delta A)^2(\Delta B)^2 \leq 0. \tag{3.12}$$

Thus we have the uncertainty relation:

$$\Delta A \cdot \Delta B \geqq \frac{\hbar}{2}. \tag{3.13}$$

---

The uncertainty relation is deeply related to the problem of measurement in quantum mechanics. Equation (3.13) quantitatively shows that a large fluctuation arises in the momentum if one determines the position accurately. Alternately, large fluctuation arises in the position if one tries to measure the momentum very precisely. The relation (3.13) implies that if one tries to determine one of the observables, position or momentum, the measurement disturbs the system so that the other observable can not be determined accurately. As can be understood from the smallness of the Planck constant $\hbar = 1.06 \times 10^{-34}$J·s, this uncertainty relation is a unique feature of the microscopic quantum world. The problem of measurement by the uncertainty relation is also deeply related to the intrinsic properties of quantum information theory.

## 3.3 Quantum states of spin $\frac{1}{2}$ particles

Particles like electrons, protons and neutrons have a quantum property called spin. Spin could be taken as an angular momentum related to the rotation of a particle on its own axis. However, an electron, for example, is considered to be a mass point without size even in the microscopic world. Since it is hard to imagine the rotation of a point, it is more natural to consider the spin as a quantum number related to an intrinsic degree of freedom of the particle. In the microscopic world, the magnitude of the orbital angular momentum is given by an integer $l$ which takes on discrete values $l = n\hbar$ $(n = 0, 1, 2, \ldots)$ in units of the Planck constant $\hbar$. The spin $s$, on the other hand, can be integer or half integer depending on the particle. In particular, fundamental particles constituting material such as electrons, protons and neutrons all have $s = \frac{1}{2}\hbar$. (From now on, we will neglect writing $\hbar$, e.g., we write $\frac{1}{2}\hbar$ as $\frac{1}{2}$. The notation is called the "**natural unit**" and commonly used in high energy physics and nuclear physics.) In general, the angular momentum $l$ has $(2l + 1)$ components. This degree of freedom is distinguished by the projection $m$ along the $z$-axis of angular momentum, and $m$ can take only the discrete values differing by one:

$$m = -l, -l + 1, \ldots, l - 1, l. \tag{3.14}$$

Let us consider a particle with spin $\frac{1}{2}$. The spin degree of freedom is fully quantum mechanical concept and its wave function can be expressed by a matrix representation. The spin 1/2 state has two degree of freedom; spin up $s_z = +1/2$, spin down $s_z = -1/2$ states. The wave function can be expressed as a state vector of two components. Let us denote these two states as the spin-up state $|\uparrow\rangle$ and the spin-down state $|\downarrow\rangle$ and express by the column vectors,

$$|\uparrow\rangle = \begin{pmatrix} 1 \\ 0 \end{pmatrix}, \qquad |\downarrow\rangle = \begin{pmatrix} 0 \\ 1 \end{pmatrix}. \tag{3.15}$$

The spin operator is defined as

$$\mathbf{s} = \frac{1}{2}\boldsymbol{\sigma} \tag{3.16}$$

where the right-hand side can be represented by $2 \times 2$ matrices:

$$\sigma_x = \begin{pmatrix} 0 & 1 \\ 1 & 0 \end{pmatrix}, \qquad \sigma_y = \begin{pmatrix} 0 & -i \\ i & 0 \end{pmatrix}, \qquad \sigma_z = \begin{pmatrix} 1 & 0 \\ 0 & -1 \end{pmatrix}. \tag{3.17}$$

These matrices were introduced by a German physicist Wolfgang Pauli, and are called the Pauli **spin matrices** after his name. They satisfy the relations:

$$\sigma_x\sigma_y = -\sigma_y\sigma_x = i\sigma_z, \quad \sigma_y\sigma_z = -\sigma_z\sigma_y = i\sigma_x, \quad \sigma_z\sigma_x = -\sigma_x\sigma_z = i\sigma_y \tag{3.18}$$

among them, and satisfy the commutation relations:

$$[\sigma_x, \sigma_y] = 2i\sigma_z, \qquad [\sigma_y, \sigma_z] = 2i\sigma_x, \qquad [\sigma_z, \sigma_x] = 2i\sigma_y, \tag{3.19}$$

the anti-commutation relations:

$$\{\sigma_i, \sigma_j\} = \sigma_i\sigma_j + \sigma_j\sigma_i = 0 \qquad (\text{if } i \neq j), \tag{3.20}$$

and the relation:

$$\sigma_i^2 = \begin{pmatrix} 1 & 0 \\ 0 & 1 \end{pmatrix} = \mathbf{1}. \tag{3.21}$$

The commutation relations (3.19) can be also expressed as

$$[s_x, s_y] = is_z, \qquad [s_y, s_z] = is_x, \qquad [s_z, s_x] = is_y. \tag{3.22}$$

Furthermore, noting that the spin-up and spin-down vector are written as (3.15), the products of ket- and bra-vectors are written as the $2\times 2$ matrices:

$$\begin{aligned} |\uparrow\rangle\langle\uparrow| &= \begin{pmatrix} 1 \\ 0 \end{pmatrix} (1\ 0) = \begin{pmatrix} 1 & 0 \\ 0 & 0 \end{pmatrix}, \\ |\uparrow\rangle\langle\downarrow| &= \begin{pmatrix} 1 \\ 0 \end{pmatrix} (0\ 1) = \begin{pmatrix} 0 & 1 \\ 0 & 0 \end{pmatrix}, \\ |\downarrow\rangle\langle\uparrow| &= \begin{pmatrix} 0 \\ 1 \end{pmatrix} (1\ 0) = \begin{pmatrix} 0 & 0 \\ 1 & 0 \end{pmatrix}, \\ |\downarrow\rangle\langle\downarrow| &= \begin{pmatrix} 0 \\ 1 \end{pmatrix} (0\ 1) = \begin{pmatrix} 0 & 0 \\ 0 & 1 \end{pmatrix}. \end{aligned} \tag{3.23}$$

Then, we can write the spin matrices in (3.17) as the products of ket- and bra-vectors:

$$\begin{aligned}\sigma_x &= |\uparrow\rangle\langle\downarrow| + |\downarrow\rangle\langle\uparrow|,\\ \sigma_y &= -i\,(|\uparrow\rangle\langle\downarrow| - |\downarrow\rangle\langle\uparrow|),\\ \sigma_z &= (|\uparrow\rangle\langle\uparrow| - |\downarrow\rangle\langle\downarrow|).\end{aligned} \tag{3.24}$$

We also note that

$$\begin{aligned} s_z|\uparrow\rangle &= \frac{1}{2}\begin{pmatrix}1 & 0\\ 0 & -1\end{pmatrix}\begin{pmatrix}1\\0\end{pmatrix} = \frac{1}{2}\begin{pmatrix}1\\0\end{pmatrix}\\ s_z|\downarrow\rangle &= \frac{1}{2}\begin{pmatrix}1 & 0\\ 0 & -1\end{pmatrix}\begin{pmatrix}0\\1\end{pmatrix} = -\frac{1}{2}\begin{pmatrix}0\\1\end{pmatrix}.\end{aligned} \tag{3.25}$$

Namely, the ket vector $|\uparrow\rangle$ points the positive direction of the $z$-axis while the vector $|\downarrow\rangle$ points the negative direction. A spin vector is in general written as

$$|\psi\rangle = a|\uparrow\rangle + b|\downarrow\rangle = \begin{pmatrix}a\\b\end{pmatrix}, \tag{3.26}$$

where the spin-up and the spin-down states are observed with the probabilities $|a|^2$ and $|b|^2$, respectively.

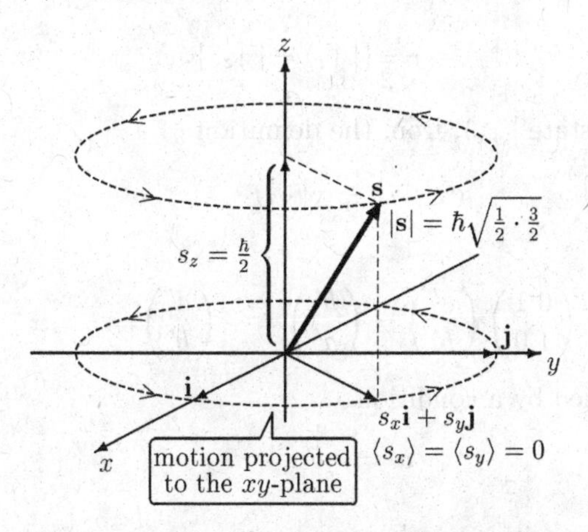


Fig. 3.2 Spin-up state $|\uparrow\rangle$ of a particle with spin $\frac{1}{2}\hbar$. Symbols $\mathbf{i}$ and $\mathbf{j}$ denote unit vectors in the directions of the $x$- and $y$-axes, respectively.

---

**Example 3.3** Express the eigenstates of $s_x$ in terms of the eigenstates $|\uparrow\rangle$ and $|\downarrow\rangle$ of $s_z$.

**Solution** Let us denote the eigenstates $+\frac{1}{2}$ and $-\frac{1}{2}$ of $s_x$ as $|\uparrow_x\rangle$ and $|\downarrow_x\rangle$, and denote the eigenstates Eq. (3.15) of $s_z$ as $|\uparrow_z\rangle, |\downarrow_z\rangle$. If we write $|\uparrow_x\rangle$ as

$$|\uparrow_x\rangle = a|\uparrow_z\rangle + b|\downarrow_z\rangle = \begin{pmatrix} a \\ b \end{pmatrix}, \tag{3.27}$$

we have

$$\langle_x\uparrow|\uparrow_x\rangle = |a|^2 + |b|^2 = 1. \tag{3.28}$$

from the normalization condition. From the definition

$$\sigma_x|\uparrow_x\rangle = |\uparrow_x\rangle, \tag{3.29}$$

we get

$$\begin{pmatrix} 0 & 1 \\ 1 & 0 \end{pmatrix}\begin{pmatrix} a \\ b \end{pmatrix} = \begin{pmatrix} b \\ a \end{pmatrix} = \begin{pmatrix} a \\ b \end{pmatrix}, \tag{3.30}$$

which is satisfied by

$$a = b. \tag{3.31}$$

If $a$ is taken to be a positive number, the state $|\uparrow_x\rangle$ is expressed as

$$|\uparrow_x\rangle = \frac{1}{\sqrt{2}}\{|\uparrow_z\rangle + |\downarrow_z\rangle\}. \tag{3.32}$$

For the spin-down state $|\downarrow_x\rangle$, from the definition

$$\sigma_x|\downarrow_x\rangle = -|\downarrow_x\rangle, \tag{3.33}$$

we obtain

$$\begin{pmatrix} 0 & 1 \\ 1 & 0 \end{pmatrix}\begin{pmatrix} a' \\ b' \end{pmatrix} = \begin{pmatrix} b' \\ a' \end{pmatrix} = -\begin{pmatrix} a' \\ b' \end{pmatrix}, \tag{3.34}$$

which can be satisfied by a condition

$$a' = -b'. \tag{3.35}$$

By choosing $a'$ as a positive number, we have

$$|\downarrow_x\rangle = \frac{1}{\sqrt{2}}\{|\uparrow_z\rangle - |\downarrow_z\rangle\}. \tag{3.36}$$

Note that Eqs. (3.32) and (3.36)) satisfy the orthogonalization condition:

$$\langle_x\uparrow|\downarrow_x\rangle = 0. \tag{3.37}$$

In a similar way, we can also obtain the eigenstates of $s_y$ for $\pm\frac{1}{2}$:

$$|\uparrow_y\rangle = \frac{1}{\sqrt{2}}\{|\uparrow_z\rangle + i|\downarrow_z\rangle\}$$
$$|\downarrow_y\rangle = \frac{1}{\sqrt{2}}\{|\uparrow_z\rangle - i|\downarrow_z\rangle\}. \tag{3.38}$$

These eigenstates of spins along the $x$- and $y$-axes can also be expressed by using the spin-rotation operator (3.78) as

$$|\uparrow_x\rangle = D_y^s\left(\frac{\pi}{2}\right)|\uparrow_z\rangle, \qquad |\downarrow_x\rangle = D_y^s\left(-\frac{\pi}{2}\right)|\uparrow_z\rangle, \tag{3.39}$$

$$|\uparrow_y\rangle = D_x^s\left(-\frac{\pi}{2}\right)|\uparrow_z\rangle, \qquad |\downarrow_y\rangle = D_x^s\left(\frac{\pi}{2}\right)|\uparrow_z\rangle. \tag{3.40}$$

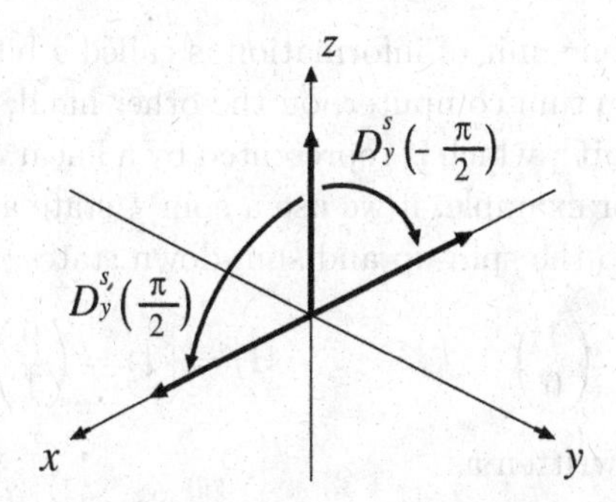


Fig. 3.3 Rotation of the eigenstate $|\uparrow_z\rangle$ of spin in the direction of the $z$-axis around the $y$-axis by angles $\pm\frac{\pi}{2}$.

**Example 3.4** Vectors **a** and **b** commute with $\boldsymbol{\sigma}$. Prove

$$(\boldsymbol{\sigma}\cdot\mathbf{a})(\boldsymbol{\sigma}\cdot\mathbf{b}) = (\mathbf{a}\cdot\mathbf{b}) + i(\mathbf{a}\times\mathbf{b})\cdot\boldsymbol{\sigma}. \tag{3.41}$$

**Solution** Equation (3.41) can be derived as follows:

$$
\begin{aligned}
&(\boldsymbol{\sigma}\cdot\mathbf{a})(\boldsymbol{\sigma}\cdot\mathbf{b}) \\
&= \sum_{i,j=(x,y,z)} \sigma_i a_i \sigma_j b_j \\
&= \sum_{i=(x,y,z)} \sigma_i^2 a_i b_i + \sum_{i\neq j} \sigma_i\sigma_j a_i b_j \\
&= \sum_{i=(x,y,z)} a_i b_i + \sigma_x\sigma_y(a_x b_y - a_y b_x) + \sigma_z\sigma_x(a_z b_x - a_x b_z) + \sigma_y\sigma_z(a_y b_z - a_z b_y) \\
&= (\mathbf{a}\cdot\mathbf{b}) + i\sigma_z[\mathbf{a}\times\mathbf{b}]_z + i\sigma_y[\mathbf{a}\times\mathbf{b}]_y + i\sigma_x[\mathbf{a}\times\mathbf{b}]_x,
\end{aligned}
$$

where the relations (3.18) $\sigma_x\sigma_y = -\sigma_y\sigma_x = i\sigma_z$, $\sigma_z\sigma_x = -\sigma_x\sigma_z = i\sigma_y$, $\sigma_y\sigma_z = -\sigma_z\sigma_y = i\sigma_x$ have been used. We have assumed that **a**, **b** are commutable with $\boldsymbol{\sigma}$, but note that the operators **a** and **b** need not commute with each other for (3.41).

---

## 3.4 Quantum bit

In classical computers, one unit of information is called a bit, and is represented by 0 or 1. In quantum computer, on the other hand, it corresponds to a **quantum bit (qubit)**, which is represented by a linear combination of two quantum states. For example, if we use a spin $\frac{1}{2}$ state as the quantum bit, we assign 0 and 1 to the spin-up and spin-down states, respectively,

$$|0\rangle = |\uparrow\rangle = \begin{pmatrix}1\\0\end{pmatrix}, \qquad |1\rangle = |\downarrow\rangle = \begin{pmatrix}0\\1\end{pmatrix}. \tag{3.42}$$

Then a quantum bit is written as

$$|\psi\rangle = a|0\rangle + b|1\rangle. \tag{3.43}$$

Equation (3.43) shows a single-qubit state. A $n$th-fold qubit state is constructed as a direct product of $n$ single-qubit states (3.43):

$$|\Psi\rangle = |\psi\rangle_1|\psi\rangle_2\cdots|\psi\rangle_n = |1,2,\ldots,n\rangle. \tag{3.44}$$

The quantum bit is also implemented by using the polarization of photon. The **linear polarization** of photon is distinguished by the **vertical polarization** and the **horizontal polarization**, which can be assigned to 0 and 1, respectively;

$$|0\rangle = |\updownarrow\rangle, \qquad |1\rangle = |\leftrightarrow\rangle. \tag{3.45}$$

A superposition of the two linearly polarized states

$$|\psi\rangle = a|\updownarrow\rangle + b|\leftrightarrow\rangle, \tag{3.46}$$

will provide various photon states such as the diagonally polarized states in directions of $+45^\circ$ and $-45^\circ$,

$$|\nearrow\rangle = \frac{1}{\sqrt{2}}\{|\updownarrow\rangle + |\leftrightarrow\rangle\} \tag{3.47}$$

$$|\nwarrow\rangle = \frac{1}{\sqrt{2}}\{|\updownarrow\rangle - |\leftrightarrow\rangle\}. \tag{3.48}$$

One also has the circularly polarized states, the **clockwise circular polarization**

$$|R\rangle = \frac{1}{\sqrt{2}}\{|\updownarrow\rangle + i|\leftrightarrow\rangle\} \tag{3.49}$$

as well as the **counterclockwise circular polarization**

$$|L\rangle = \frac{1}{\sqrt{2}}\{|\updownarrow\rangle - i|\leftrightarrow\rangle\}. \tag{3.50}$$

These states are also adopted as quantum bits. For the **circular polarizations** of Eqs. (3.49) and (3.50), the angular momentum is quantized along the direction of the photon propagation. The clockwise polarized state (3.49) has the value $-1$ for the projected component ( $h = -1$) while the counterclockwise state of Eq. (3.50) has + (helicity $h = +1$) (See Exercise [6] of Chapter 3).

## 3.5 Angular momentum, spin and rotation

Let us consider the rotation of a wave function $\psi(x, y, z)$ in the three-dimensional space. When we consider the rotation of wave function, we have two choices: one is the rotation of the coordinate axes while the other is the rotation of wave function itself. Here we adopt the latter, i.e., the rotation of the wave function. Let us consider a vector in the $xy$ plane $\mathbf{r} = (x, y, 0) = (r\cos\alpha_0, r\sin\alpha_0, 0)$, where $\alpha_0$ is the angle between the $x$-axis and $\mathbf{r}$. Rotate $\mathbf{r}$ about the $z$-axis by an angle $\alpha$. Since the vector $\mathbf{r}(x, y, z)$ is changed into

$$\begin{aligned}\mathbf{r}' &= (r\cos(\alpha + \alpha_0), r\sin(\alpha + \alpha_0), 0)\\ &= (\cos\alpha x - \sin\alpha y, \sin\alpha x + \cos\alpha y, 0),\end{aligned} \tag{3.51}$$

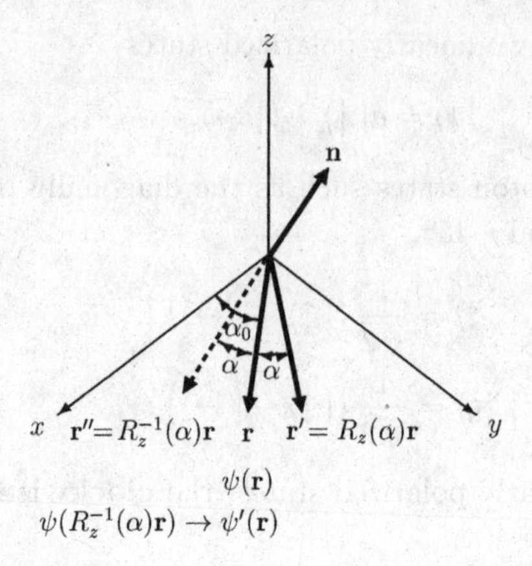


Fig. 3.4 The rotation of position vector $\mathbf{r}$ and the rotation of wave function $\psi(\mathbf{r})$. Symbol $\mathbf{n}$ means a unit vector in an arbitrary direction.

we can express the rotation in a matrix form for the $\mathbf{r}' = (x', y', z')$ as

$$\begin{pmatrix} x' \\ y' \\ z' \end{pmatrix} = \begin{pmatrix} \cos\alpha & -\sin\alpha & 0 \\ \sin\alpha & \cos\alpha & 0 \\ 0 & 0 & 1 \end{pmatrix} \begin{pmatrix} x \\ y \\ z \end{pmatrix}. \tag{3.52}$$

Thus the rotation about the $z$-axis by an angle $\alpha$ can be given by a matrix

$$R_z(\alpha) = \begin{pmatrix} \cos\alpha & -\sin\alpha & 0 \\ \sin\alpha & \cos\alpha & 0 \\ 0 & 0 & 1 \end{pmatrix}. \tag{3.53}$$

and its Hermitian conjugate is

$$R_z^\dagger(\alpha) = \begin{pmatrix} \cos\alpha & \sin\alpha & 0 \\ -\sin\alpha & \cos\alpha & 0 \\ 0 & 0 & 1 \end{pmatrix}. \tag{3.54}$$

The matrices (3.53) and (3.54) satisfy the unitarity condition:

$$R_z^\dagger(\alpha)R_z(\alpha) = \begin{pmatrix} 1 & 0 & 0 \\ 0 & 1 & 0 \\ 0 & 0 & 1 \end{pmatrix} = \mathbf{1}, \tag{3.55}$$

namely, $R_z(\alpha)$ is a unitary operator. Similarly, the rotations about the $x$- and the $y$-axes can be represented by the matrices:

$$R_x(\alpha) = \begin{pmatrix} 1 & 0 & 0 \\ 0 & \cos\alpha & -\sin\alpha \\ 0 & \sin\alpha & \cos\alpha \end{pmatrix} \tag{3.56}$$

$$R_y(\alpha) = \begin{pmatrix} \cos\alpha & 0 & \sin\alpha \\ 0 & 1 & 0 \\ -\sin\alpha & 0 & \cos\alpha \end{pmatrix}. \tag{3.57}$$

The successive rotations by $\alpha_x$ about the $x$-axis, by $\alpha_y$ the $y$-axis and by $\alpha_z$ about the $z$-axis are realized by the operator

$$R(\alpha_x, \alpha_y, \alpha_z) = R_z(\alpha_z)R_y(\alpha_y)R_x(\alpha_x) \tag{3.58}$$

applying to the vector $\mathbf{r}$.

The rotation of wave function $\psi(\mathbf{r})$ should be distinguished strictly from that of the coordinates. Let us denote the rotation of the wave function as

$$\psi'(\mathbf{r}) = D(\alpha)\psi(\mathbf{r}) \tag{3.59}$$

where $D(\alpha)$ is an operator to act on the wave function. Since the norm of wave function is invariant under rotation, we have

$$\langle\psi'|\psi'\rangle = \langle\psi|D^\dagger(\alpha)D(\alpha)|\psi\rangle = \langle\psi|\psi\rangle, \tag{3.60}$$

which provides the unitary condition for the rotation operator,

$$D^\dagger(\alpha)D(\alpha) = 1. \tag{3.61}$$

When the wave function rotates by an angle $\alpha$ about the $z$-axis, the function is changed by the transformation $\mathbf{r}' = R_z(\alpha)\mathbf{r}$. This means that the function at the position $\mathbf{r}'' = R_z^{-1}(\alpha)\mathbf{r}$ is carried to the new position $\mathbf{r}$, i.e.,

$$\psi'(\mathbf{r}) = D_z(\alpha)\psi(\mathbf{r}) = \psi(R_z^{-1}(\alpha)\mathbf{r}). \tag{3.62}$$

Note that the rotation of the coordinates is an action opposite to the rotation of the wave function with respect to the position $\mathbf{r}$. Then, (3.62) can be expressed as

$$\psi'(\mathbf{r}) = \psi(x\cos\alpha + y\sin\alpha, -x\sin\alpha + y\cos\alpha, z). \tag{3.63}$$

Replacing $\alpha$ by an infinitesimal rotation $d\alpha$ and applying the Taylor expansion to the first order in $d\alpha$, Eq. (3.62) is approximated as

$$\begin{aligned} D_z(d\alpha)\psi(x,y,z) &\approx \psi(x + yd\alpha, -xd\alpha + y, z) \\ &\approx \psi(x,y,z) + \left(y\frac{\partial\psi}{\partial x} - x\frac{\partial\psi}{\partial y}\right)d\alpha \\ &= (1 - il_z d\alpha)\psi, \end{aligned} \tag{3.64}$$

where $l_z$ is the $z$-component of the **orbital angular momentum operator**, given by

$$l_z = xp_y - yp_x = -i\left(x\frac{\partial}{\partial y} - y\frac{\partial}{\partial x}\right), \tag{3.65}$$

(the convention $\hbar = 1$ is taken). Similarly, the $x$- and $y$-components of the orbital angular momentum operator are given by

$$l_x = yp_z - zp_y = -i\left(y\frac{\partial}{\partial z} - z\frac{\partial}{\partial y}\right), \tag{3.66}$$

$$l_y = zp_x - xp_z = -i\left(z\frac{\partial}{\partial x} - x\frac{\partial}{\partial z}\right). \tag{3.67}$$

The angular momentum operators satisfy commutation relations

$$[l_x, l_y] = il_z, \quad [l_y, l_z] = il_x, \quad [l_z, l_x] = il_y. \tag{3.68}$$

Note that Eqs. (3.22) for the spin operators $\mathbf{s}$ satisfy the similar relations. Going back to (3.64), we have obtained the operator for infinitesimal rotation as

$$D_z(d\alpha) = 1 - il_z d\alpha. \tag{3.69}$$

The finite rotation can be obtained by successive infinitesimal rotations as

$$D_z(\alpha + d\alpha) = D_z(d\alpha)D_z(\alpha). \tag{3.70}$$

Inserting (3.69) into (3.70), we have

$$D_z(\alpha + d\alpha) = (1 - il_z d\alpha)D_z(\alpha), \tag{3.71}$$

Equation (3.70) turns out to be a differential equation

$$\frac{D_z(\alpha + d\alpha) - D_z(\alpha)}{d\alpha} = \frac{dD_z(\alpha)}{d\alpha} = -il_z D_z(\alpha). \tag{3.72}$$

Integrating (3.72) with the condition $D_z(0) = 1$, we obtain

$$D_z(\alpha) = e^{-il_z\alpha}D_z(0) = e^{-il_z\alpha}. \tag{3.73}$$

This equation can be generalized to the rotation by an angle $\alpha$ around an arbitrary direction given by a unit vector $\hat{\mathbf{n}}$ in Fig. 3.4,

$$D(\alpha) = e^{-i\mathbf{l}\cdot\hat{\mathbf{n}}\alpha}. \tag{3.74}$$

Since the spin angular momentum operators have the same relations as the orbital angular momentum operators, we expect that the rotation operator about $i$-axis is

$$D_i^s(\alpha) = e^{-i\sigma_i\alpha/2}. \tag{3.75}$$

Noting the even and the odd powers of the spin function $\sigma_i$

$$\sigma_i^{2n} = \mathbf{1}, \qquad \sigma_i^{2n+1} = \sigma_i, \tag{3.76}$$

and the power-series expansion of the exponential function,

$$e^{-ix} = \sum_{n=0}^{\infty} \frac{(-ix)^n}{n!} = \sum_{n=0}^{\infty} (-)^n \frac{x^{2n}}{(2n)!} - i \sum_{n=0}^{\infty} (-)^n \frac{x^{2n+1}}{(2n+1)!}, \tag{3.77}$$

We have an expression of Eq. (3.75) as

$$D_i^s(\alpha) = \cos\left(\frac{\alpha}{2}\right) \mathbf{1} - i \sin\left(\frac{\alpha}{2}\right) \sigma_i. \tag{3.78}$$

We will now verify that this operator actually rotates a spin $\frac{1}{2}$ state by $\alpha$ about the $z$-axis. Let us write the state after the rotation as

$$|\psi'\rangle = D_z^s(\alpha)|\psi\rangle. \tag{3.79}$$

This rotation will change the expectation value of spin operator $\sigma_x$ in the state $|\psi\rangle$ into

$$\langle\psi|\sigma_x|\psi\rangle \longrightarrow \langle\psi'|\sigma_x|\psi'\rangle = \langle\psi|e^{i\sigma_z\alpha/2}\sigma_x e^{-i\sigma_z\alpha/2}|\psi\rangle. \tag{3.80}$$

Rewriting $\sigma_x$ in (3.80) by the tensor product of the bra and ket vectors in (3.24), we can derive

$$\begin{aligned} e^{i\sigma_z\alpha/2}\sigma_x e^{-i\sigma_z\alpha/2} &= e^{i\sigma_z\alpha/2}\{|\uparrow\rangle\langle\downarrow| + |\downarrow\rangle\langle\uparrow|\}e^{-i\sigma_z\alpha/2} \\ &= e^{i\alpha/2}|\uparrow\rangle\langle\downarrow|e^{i\alpha/2} + e^{-i\alpha/2}|\downarrow\rangle\langle\uparrow|e^{-i\alpha/2} \\ &= \cos\alpha(|\uparrow\rangle\langle\downarrow| + |\downarrow\rangle\langle\uparrow|) + i\sin\alpha(|\uparrow\rangle\langle\downarrow| - |\downarrow\rangle\langle\uparrow|) \\ &= \cos\alpha\sigma_x - \sin\alpha\sigma_y. \end{aligned} \tag{3.81}$$

Thus the expectation value (3.80) is expressed as

$$\langle\psi'|\sigma_x|\psi'\rangle = \cos\alpha\langle\psi|\sigma_x|\psi\rangle - \sin\alpha\langle\psi|\sigma_y|\psi\rangle. \tag{3.82}$$

Equation (3.82) shows that the expectation value of $\sigma_x$ is rotated about the $z$-axis by an angle $\alpha$ due to the rotation of spin $\frac{1}{2}$ state. If we put $\alpha = \frac{\pi}{2}$, (3.82) becomes

$$\langle\psi'|\sigma_x|\psi'\rangle = -\langle\psi|\sigma_y|\psi\rangle.$$

This result shows that the expectation value of $-\sigma_y$ by $|\psi\rangle$ is equivalent to the expectation value of $\sigma_x$ by $|\psi'\rangle$ after the rotation of state $|\psi\rangle$. This fact is understood as the change of the direction of spin by the rotation of state, as seen Fig. 3.5. On the other hand, since $[\sigma_z, e^{-i\sigma_z\alpha/2}] = 0$, the expectation value of $\sigma_z$ is unchanged;

$$\langle\psi'|\sigma_z|\psi'\rangle = \langle\psi|e^{i\sigma_z\alpha/2}\sigma_z e^{-i\sigma_z\alpha/2}|\psi\rangle = \langle\psi|\sigma_z|\psi\rangle. \tag{3.83}$$

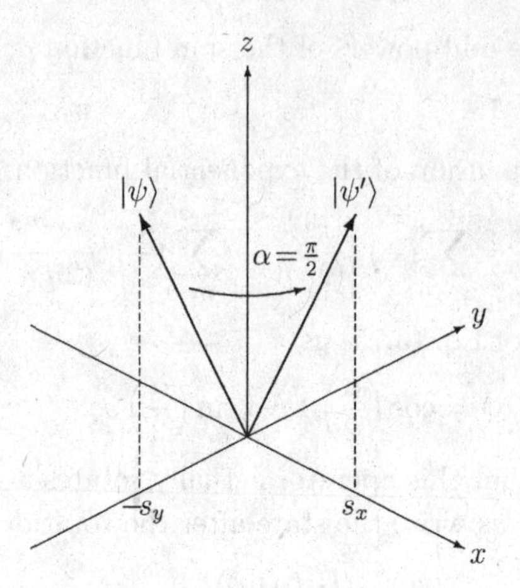


Fig. 3.5 Rotation of the state $|\psi\rangle$ about $z$-axis by an angle $\alpha = \frac{\pi}{2}$.

In this way, we have seen that the operator $D_z^s(\alpha)$ rotates the state $|\psi\rangle$ by $\alpha$ about the $z$-axis.

A remarkable point with the spin $\frac{1}{2}$ state is that the state does not return to the original state by the rotation by $\alpha = 2\pi$. In fact, let us write a ket vector $|\psi\rangle$ as

$$|\psi\rangle = |\uparrow\rangle\langle\uparrow|\psi\rangle + |\downarrow\rangle\langle\downarrow|\psi\rangle. \tag{3.84}$$

using

$$\begin{pmatrix} 1 & 0 \\ 0 & 1 \end{pmatrix} = |\uparrow\rangle\langle\uparrow| + |\downarrow\rangle\langle\downarrow|, \tag{3.85}$$

By applying $D_z^s(2\pi)$ to this state, we have

$$\begin{aligned} |\psi'\rangle &= D_z^s(2\pi)|\psi\rangle = e^{-i\sigma_z\pi}|\uparrow\rangle\langle\uparrow|\psi\rangle + e^{-i\sigma_z\pi}|\downarrow\rangle\langle\downarrow|\psi\rangle \\ &= -(|\uparrow\rangle\langle\uparrow|\psi\rangle + |\downarrow\rangle\langle\downarrow|\psi\rangle) = -|\psi\rangle. \end{aligned} \tag{3.86}$$

Namely, the rotation by the angle $2\pi$ gives a phase $(-)$ so that the rotation by $4\pi$ would be needed to return to the original state. This fact is well known as a character of the rotation of a spin $\frac{1}{2}$ state.

Note that Eq. (3.69) has the form equivalent to the time-evolution operator $U(\Delta t)$ of Eq. (2.25). To be more specific, we will obtain Eq. (3.69) if we replace Hamiltonian $H$ by $l_z$ and $\Delta t$ by $d\alpha$ of Eq. (2.25). This similarity is a general character that holds also in the operator of Galilean transformation.

**Summary of Chapter 3**

(1) The operators $\mathbf{r} = (x, y, z)$ and $\mathbf{p} = (p_x.p_y, p_z)$ satisfy the commutation relations;

$$[x, p_x] = i\hbar, \ \ [y, p_y] = i\hbar, \ \ [z, p_z] = i\hbar.$$

(2) If the Hermitian operators $A$ and $B$ satisfy the commutation relation $[A, B] = i\hbar$, the uncertainty relation

$$\Delta A \cdot \Delta B \geq \hbar/2$$

holds.

(3) The spin 1/2 states are expressed by column vectors and used as a set of basis states of qubits with bit values 0 and 1;

$$|\uparrow\rangle = \begin{pmatrix} 1 \\ 0 \end{pmatrix} = |0\rangle, \qquad |\downarrow\rangle = \begin{pmatrix} 0 \\ 1 \end{pmatrix} = |1\rangle.$$

(4) The qubit is implemented by a linear combination of the two spin states as,

$$|\psi\rangle = a|0\rangle + b|1\rangle.$$

## 3.6 Problems

[1] Express a linear polarization state of a photon pair

$$|\Psi\rangle_{12} = \frac{1}{\sqrt{2}} \{|\updownarrow\rangle_1 |\updownarrow\rangle_2 - |\leftrightarrow\rangle_1 |\leftrightarrow\rangle_2\} \tag{3.87}$$

by circular polarization states $|R\rangle$ and $|L\rangle$.

[2] Express the eigenstates of spin $s_y$ by the eigenstates of $s_z$, $|\uparrow_z\rangle$ and $|\downarrow_z\rangle$.

[3] Show that $\mathbf{s}^2$ commute with $s_z$.

[4] Show that the spin operators

$$\sigma_x = |\uparrow\rangle\langle\downarrow| + |\downarrow\rangle\langle\uparrow|,$$
$$\sigma_y = -i\left(|\uparrow\rangle\langle\downarrow| - |\downarrow\rangle\langle\uparrow|\right),$$
$$\sigma_z = \left(|\uparrow\rangle\langle\uparrow| - |\downarrow\rangle\langle\downarrow|\right)$$

satisfy the commutation relations in Eqs. (3.19) and (3.20).

[5] Show that the orbital angular momentum operators $l_x, l_y$ and , $l_z$ satisfy the commutation relations given by Eq. (3.68).

[6] Rotate the circular polarization states in the $xy$-plane

$$|L\rangle = \frac{1}{\sqrt{2}}\{|\updownarrow\rangle - i|\leftrightarrow\rangle\}$$
$$|R\rangle = \frac{1}{\sqrt{2}}\{|\updownarrow\rangle + i|\leftrightarrow\rangle\}$$

around the $z$-axis by an angle $d\alpha$ ($d\alpha \ll 1$) and show that these states have the angular momentum components $l_z = \pm 1$ along the $z$-axis (helicity $h = (\mathbf{p} \cdot \mathbf{l})/\mathrm{p} = \pm 1$.) using the operator (3.74).

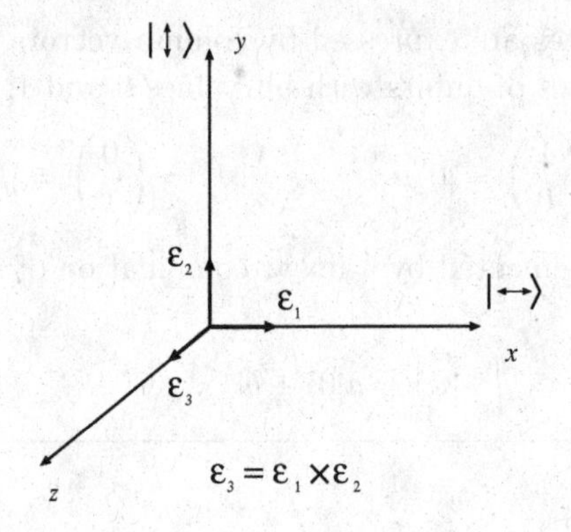


Fig. 3.6 Unit vectors $\varepsilon_1, \varepsilon_2, \varepsilon_3$ in the directions of $x, y, z$ and the linear polarized states $|\leftrightarrow\rangle, |\updownarrow\rangle$.

Chapter 4

# EPR Pair and Measurement

In the years of 1920s when the quantum mechanics was being established, there was a famous controversy between Bohr and Einstein about the statistical interpretation of quantum mechanics. The controversy was finally concluded by an emphatic victory of Bohr, since his statistical interpretation and the duality of wave and particle explained many experimental facts without any difficulty. The "Gedanken" experiment on two entangled particles (the EPR pair) presented by Einstein, Podolski and Rosen in the course of the controversy, however, has had a continuous influence for many years on quantum communication and quantum cryptography which are being developed at the present day.

**Keywords: entanglement; EPR pair; Bell's inequality; hidden variable; GHZ states.**

## 4.1 EPR pair

An EPR pair is composed of a superposition of two particles with spin $\frac{1}{2}$, or two polarized photons. The abbreviation "EPR" came from initials of Einstein, Podolski and Rosen who first proposed the idea of "entanglement" of particles. Here we consider the Gedanken experiment invented by D. Bohm. Let us suppose that two particles of spin $\frac{1}{2}$ couple to the total spin 0. The particles can be electrons or protons, or any similar particles. The wave function of total spin $S = 0$, where $\boldsymbol{S} = \boldsymbol{s}_1 + \boldsymbol{s}_2$, is expressed by a superposition of spin-up and spin-down states:

$$|S=0\rangle = \frac{1}{\sqrt{2}}\{|\uparrow\rangle_1|\downarrow\rangle_2 - |\downarrow\rangle_1|\uparrow\rangle_2\}. \tag{4.1}$$

**Example 4.1** Show that the two-particle state (4.1) has total spin

$S = 0$.

**Solution** The total spin operator of the two-particle system is expressed as

$$\mathbf{S}^2 = (\mathbf{s}_1 + \mathbf{s}_2)^2$$
$$= \frac{1}{4}(\boldsymbol{\sigma}_1^2 + \boldsymbol{\sigma}_2^2 + 2\boldsymbol{\sigma}_1 \cdot \boldsymbol{\sigma}_2) = \frac{1}{4}(3 + 3 + 2\boldsymbol{\sigma}_1 \cdot \boldsymbol{\sigma}_2) = \frac{1}{2}(3 + \boldsymbol{\sigma}_1 \cdot \boldsymbol{\sigma}_2), \tag{4.2}$$

where we make use of Eq. (3.21). The expectation value of scalar product of spin operators $\boldsymbol{\sigma}_1 \cdot \boldsymbol{\sigma}_2 = \sigma_{x1}\sigma_{x2} + \sigma_{y1}\sigma_{y2} + \sigma_{z1}\sigma_{z2}$ can be evaluated by using the following identities derived from Eq. (3.24),

$$\begin{aligned} \sigma_x|\uparrow\rangle &= |\downarrow\rangle, & \sigma_x|\downarrow\rangle &= |\uparrow\rangle, \\ \sigma_y|\uparrow\rangle &= i|\downarrow\rangle, & \sigma_y|\downarrow\rangle &= -i|\uparrow\rangle, \\ \sigma_z|\uparrow\rangle &= |\uparrow\rangle, & \sigma_z|\downarrow\rangle &= -|\downarrow\rangle. \end{aligned} \tag{4.3}$$

With these identities, we find that the non-diagonal terms of $\sigma_{x1}\sigma_{x2}, \sigma_{y1}\sigma_{y2}$ and the diagonal terms of $\sigma_{z1}\sigma_{z2}$ remain to be non-zero:

$$\begin{aligned} \langle S = 0|\boldsymbol{\sigma}_1 \cdot \boldsymbol{\sigma}_2|S = 0\rangle &= -\frac{2}{2}\, {}_1\langle\uparrow|_2\langle\downarrow|\{\sigma_{x1}\sigma_{x2} + \sigma_{y1}\sigma_{y2}\}|\downarrow\rangle_1|\uparrow\rangle_2 \\ &\quad + \frac{2}{2}\, {}_1\langle\uparrow|_2\langle\downarrow|\sigma_{z1}\sigma_{z2}|\uparrow\rangle_1|\downarrow\rangle_2 \\ &= -(1 \cdot 1 - i \cdot i) + (1 \cdot (-1)) = -3. \end{aligned} \tag{4.4}$$

so that the expectation value of Eq. (4.2) becomes $\langle S = 0|\mathbf{S}^2|S = 0\rangle = S(S+1) = 0$ as same as the total spin $S = 0$.

---

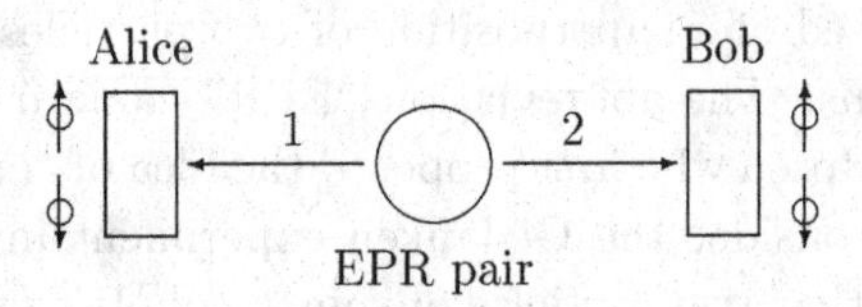


Fig. 4.1 EPR pair and directions of spins measured by Alice and Bob.

Let us suppose that Alice and Bob separately measure the spin of this $S = 0$ state at two remote places. If Alice measures $z$-component of spin labeled as 1, the probabilities of finding spin-up and spin-down states are 50% respectively because they have the same amplitudes. The same argument can be applied to Bob who measures the spin labeled as 2. If Alice finds the spin-up state, however, the spin of Bob must be down with 100% probability. If Alice is on the earth and Bob is on the moon, or even if they are in different galaxies, the remote action (so called "action at distance") propagates the correlation by a contraction of wave functions. This is the commonly accepted interpretation of quantum mechanics these days.

Einstein and his collaborators were opposed to this interpretation. They insisted the following idea: there were some unknown parameters (hidden variables) which were not measured directly by experiment. These hidden parameters make the quantum mechanical world as if ruled by the probability. This controversy on quantal measurements had been discussed for a long time as a philosophical conceptual problem until 1964 when J. S. Bell proposed an inequality (known as the Bell inequality) by which the problem could be solved experimentally. At the same time, experiment on the EPR pair gave rise to a completely new method of communications. This is called quantum communication or quantum teleportation. It is a revolutionary method of transmitting information between remote places (even between different galaxies) at an instant using entangled particles.

## 4.2 Transmission of quantum states

Let us consider how to send from Alice to Bob a quantum state

$$|\phi\rangle_1 = a|\uparrow\rangle_1 + b|\downarrow\rangle_1 = \begin{pmatrix} a \\ b \end{pmatrix} \tag{4.5}$$

using an EPR pair. We label the original quantum state by index 1, while the EPR pair has the indices 2 and 3. The wave function of EPR pair is written as

$$|\Psi\rangle_{23} = \sqrt{\frac{1}{2}}\,\{|\uparrow\rangle_2|\downarrow\rangle_3 - |\downarrow\rangle_2|\uparrow\rangle_3\}\,. \tag{4.6}$$

The particle "2" is sent to Alice, while "3" is transmitted to Bob. One should note that EPR pair has no information of the wave function (4.5) which Alice wants to send. It is possible to expand the coupled three-particle state of $\phi$ and $\Psi$ in terms of the complete set called "the Bell states":

$$|\Psi^{(\pm)}\rangle_{12} = \frac{1}{\sqrt{2}}\{|\uparrow\rangle_1|\downarrow\rangle_2 \pm |\downarrow\rangle_1|\uparrow\rangle_2\} \tag{4.7}$$

$$|\Phi^{(\pm)}\rangle_{12} = \frac{1}{\sqrt{2}}\{|\uparrow\rangle_1|\uparrow\rangle_2 \pm |\downarrow\rangle_1|\downarrow\rangle_2\}. \tag{4.8}$$

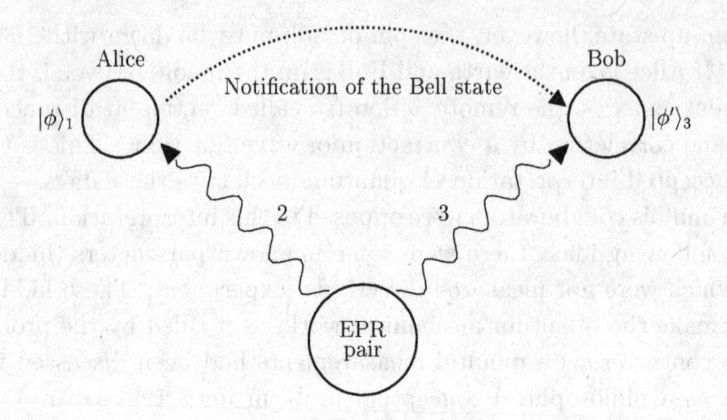


Fig. 4.2 Quantum teleportation with EPR pair.

The EPR pair (4.6) corresponds to $|\Psi^{(-)}\rangle$. A three-particle state $|\Psi\rangle_{123} = |\phi\rangle_1|\Psi\rangle_{23}$ can be expressed by these Bell states as

$$\begin{aligned}|\Psi\rangle_{123} &= |\phi\rangle_1|\Psi\rangle_{23} \\ &= \frac{1}{2}\Big[|\Psi^{(-)}\rangle_{12}(-a|\uparrow\rangle_3 - b|\downarrow\rangle_3) + |\Psi^{(+)}\rangle_{12}(-a|\uparrow\rangle_3 + b|\downarrow\rangle_3) \\ &\quad + |\Phi^{(-)}\rangle_{12}(a|\downarrow\rangle_3 + b|\uparrow\rangle_3) + |\Phi^{(+)}\rangle_{12}(a|\downarrow\rangle_3 - b|\uparrow\rangle_3)\Big]. \qquad (4.9)\end{aligned}$$

Equation (4.9) suggests that Alice measures all the Bell states with equal probabilities. The state labeled 3 which Bob observes is determined at the same instant when Alice observes one of the Bell states. Thus the state 3 of Bob has a direct relationship to the Alice's initial state, as summarized in Table 4.1. If Alice finds $\Psi_{12}^{(-)}$, for example, Bob will receive Alice's original state with an phase factor $(-1)$. In other three cases, Bob will have the state transformed by one of the matrices: $\sigma_x, \sigma_y, \sigma_z$. In this way, the quantum information will be transmitted from Alice to Bob at an instant. This is a technique called " quantum teleportation". If Alice let Bob know which Bell state she observed, Bob can reproduce the original information from Alice exactly by making a proper transformation using one of $1, \sigma_x, \sigma_y, \sigma_z$. On the other hand, the original information does not remain in the observed state by Alice so that it is kept secret during the transmission of EPR pair. In this way, quantum teleportation using EPR pair has a good potentiality to send information fast and safe. Experiment on quantum teleportation succeeded in 1990 by using polarized photon pairs to prove the transmission of information based on quantum mechanics.

Table 4.1 Relation between Alice's Bell state and Bob's state.

| Alice | Bob |
|---|---|
| $\Psi_{12}^{(-)}$ | $-a\|\uparrow\rangle_3 - b\|\downarrow\rangle_3 = -\begin{pmatrix} a \\ b \end{pmatrix}_3$ |
| $\Psi_{12}^{(+)}$ | $-a\|\uparrow\rangle_3 + b\|\downarrow\rangle_3 = \begin{pmatrix} -a \\ b \end{pmatrix}_3 = -\sigma_z \begin{pmatrix} a \\ b \end{pmatrix}_3$ |
| $\Phi_{12}^{(-)}$ | $b\|\uparrow\rangle_3 + a\|\downarrow\rangle_3 = \begin{pmatrix} b \\ a \end{pmatrix}_3 = \sigma_x \begin{pmatrix} a \\ b \end{pmatrix}_3$ |
| $\Phi_{12}^{(+)}$ | $-b\|\uparrow\rangle_3 + a\|\downarrow\rangle_3 = \begin{pmatrix} -b \\ a \end{pmatrix}_3 = -i\sigma_y \begin{pmatrix} a \\ b \end{pmatrix}_3$ |

## 4.3 Einstein's locality principle in quantum mechanics

The concept of Einstein on physical phenomena in microscopic world is based on the locality principle. It is such an idea that "Any physical phenomenon that Bob observes is independent of Alice's observation at a distant point." This is the concept underlying theory of hidden variables. Let us see how theory of hidden variables describes measurements of a particle with spin $\frac{1}{2}$. We measure the projection of spin on $z$ and $x$ axes. According

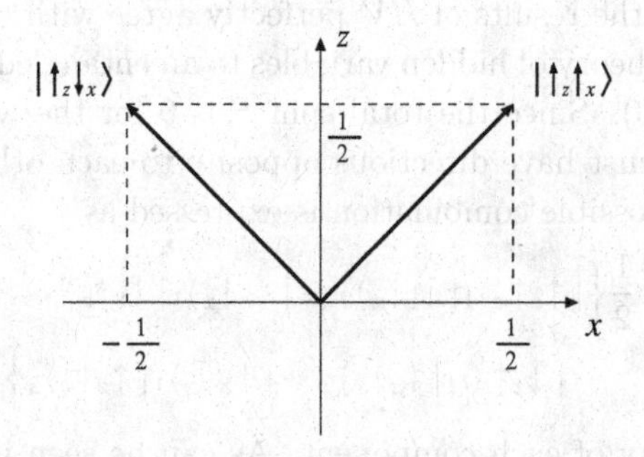


Fig. 4.3 The direction of the spin in the theory of hidden variables.

to hidden variable theory, one can measure $z$- and $x$-components of spin simultaneously. When the projection of spin on the $z$ axis is positive, the $x$-component of spin has equal probabilities for positive and negative directions. If we treat $s_x$ as a hidden variable, the state $|\uparrow_z\rangle$ is written as

$$|\uparrow_z\rangle_{HV} = \int_{-\frac{1}{2}}^{\frac{1}{2}} |\uparrow_z, \lambda\rangle w(\lambda)\, d\lambda, \tag{4.10}$$

where $HV$ means the hidden variable and $w(\lambda)$ is a weight function. If the hidden variable takes only $+1/2$ and $-1/2$ values, Equation (4.10) can be expressed as

$$|\uparrow_z\rangle_{HV} = \frac{1}{\sqrt{2}}\{|\uparrow_z\uparrow_x\rangle + |\uparrow_z\downarrow_x\rangle\}. \tag{4.11}$$

The expectation values for $\sigma_z$ and $\sigma_x$ then become

$$\langle\uparrow_z|\sigma_z|\uparrow_z\rangle_{HV} = \frac{1}{2} + \frac{1}{2} = 1 \tag{4.12}$$

$$\langle\uparrow_z|\sigma_x|\uparrow_z\rangle_{HV} = \frac{1}{2} - \frac{1}{2} = 0. \tag{4.13}$$

These results agree with those of quantum mechanics. To clarify this point, we express the wave function in quantum mechanics $|\uparrow_z\rangle_{QM}$ as

$$|\uparrow_z\rangle_{QM} = \frac{1}{\sqrt{2}}\{|\uparrow_x\rangle + |\downarrow_x\rangle\} \tag{4.14}$$

using the eigenstates of $\sigma_x$: Eq. (3.32) and Eq. (3.36). Then we have

$$\langle\uparrow_z|\sigma_z|\uparrow_z\rangle_{QM} = 1 \tag{4.15}$$

$$\langle\uparrow_z|\sigma_x|\uparrow_z\rangle_{QM} = \frac{1}{2}(1-1) = 0 \tag{4.16}$$

It is now proved that the results of $HV$ perfectly agree with those of $QM$.

Next let us apply theory of hidden variables to an entangled two-particle pair expressed by (4.6). Since the total spin $S$ is 0 for the wave function (4.6), spins 1 and 2 must have directions opposite to each other both in $z$ and $x$ directions. A possible combination is expressed as

$$\begin{aligned}|\Psi\rangle_{HV} = \frac{1}{2}\Big\{&|\uparrow_z\uparrow_x\rangle_1|\downarrow_z\downarrow_x\rangle_2 + |\uparrow_z\downarrow_x\rangle_1|\downarrow_z\uparrow_x\rangle_2 \\ &- |\downarrow_z\uparrow_x\rangle_1|\uparrow_z\downarrow_x\rangle_2 - |\downarrow_z\downarrow_x\rangle_1|\uparrow_z\uparrow_x\rangle_2\Big\}\end{aligned} \tag{4.17}$$

having 25% probability of each component. As can be seen in Eq. (4.17), if the spin of particle 1 is positive along $z$ axis, then the particle 2 has negative value, while if the particle 1 has the $z$ component of spin negative then the particle 2 has positive value. This property is in perfect agreement with the result of quantum mechanics for the state with $S = 0$. One should notice, however, that the spin $x$ component has a definite value in all the cases, regardless of whether it is observed or not.

In the case of an independent particle, theory of hidden variables gives the same result as quantum mechanics. How about the entangled two-particle state given by Eq. (4.17)? We take three unit vectors $\mathbf{a}, \mathbf{b}, \mathbf{c}$ in $xz$-plane to measure the spin orientation. Assuming the spin projection is distinguished only by the signs + or − in each direction, we will evaluate probabilities of two particles to have positive correlation ++ in the directions specified by two vectors **a**-**b**, **a**-**c** and **c**-**b**, respectively.

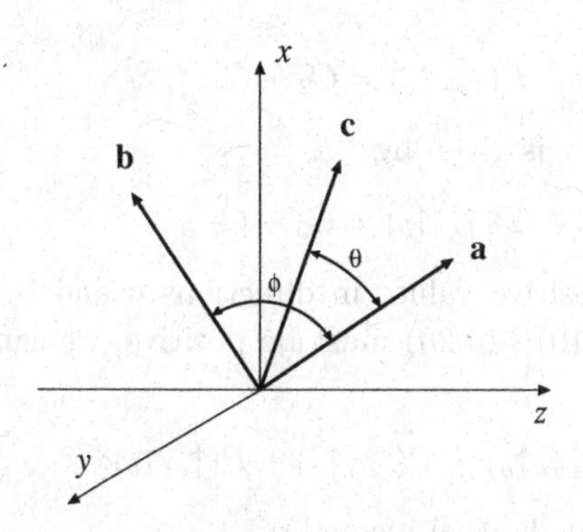


Fig. 4.4 Three detectors along vectors $\mathbf{a}, \mathbf{b}, \mathbf{c}$ in $xz$-plane.

Table 4.2 Possible combinations of spin measurements of a two-particle pair in $\mathbf{a}, \mathbf{b}, \mathbf{c}$ directions by theory of hidden variable.

| the observer | Alice | Bob |
|---|---|---|
| probability | particle 1 | particle 2 |
| $C_1$ | $\uparrow_a\uparrow_b\uparrow_c$ | $\downarrow_a\downarrow_b\downarrow_c$ |
| $C_2$ | $\uparrow_a\uparrow_b\downarrow_c$ | $\downarrow_a\downarrow_b\uparrow_c$ |
| $C_3$ | $\uparrow_a\downarrow_b\uparrow_c$ | $\downarrow_a\uparrow_b\downarrow_c$ |
| $C_4$ | $\uparrow_a\downarrow_b\downarrow_c$ | $\downarrow_a\uparrow_b\uparrow_c$ |
| $C_5$ | $\downarrow_a\uparrow_b\uparrow_c$ | $\uparrow_a\downarrow_b\downarrow_c$ |
| $C_6$ | $\downarrow_a\uparrow_b\downarrow_c$ | $\uparrow_a\downarrow_b\uparrow_c$ |
| $C_7$ | $\downarrow_a\downarrow_b\uparrow_c$ | $\uparrow_a\uparrow_b\downarrow_c$ |
| $C_8$ | $\downarrow_a\downarrow_b\downarrow_c$ | $\uparrow_a\uparrow_b\uparrow_c$ |

Table 4.2 shows possible combinations of spin measurements of a two-particle pair along vectors $\mathbf{a}, \mathbf{b}, \mathbf{c}$. For example, the row with probability $C_1$ shows the case when measurement of particle 1 along $\mathbf{a}$ is + and measurements in the directions $\mathbf{b}, \mathbf{c}$ also give +. Since the total spin is zero, the spin of particle 2 must inevitably have − values in all the directions $\mathbf{a}, \mathbf{b}$, and $\mathbf{c}$.

Results of measurements can be classified by all possible combinations in Table 4.2. Now, let use consider the case in which Alice measures the particle 1 in **a** direction while Bob measures the particle 2 in **b** direction. The probability of both having the positive value is

$$P(\uparrow_a, \uparrow_b) = C_3 + C_4. \tag{4.18}$$

For the measurement of directions **a** and **c**, the probability of both having positive value is

$$P(\uparrow_a, \uparrow_c) = C_2 + C_4. \tag{4.19}$$

Similarly, the probability is given by

$$P(\uparrow_c, \uparrow_b) = C_3 + C_7 \tag{4.20}$$

for measurements of positive values in directions **c** and **b**. Since all the probabilities (4.18), (4.19) , (4.20) must be positive, we can derive an inequality:

$$P(\uparrow_a, \uparrow_b) \leq P(\uparrow_a, \uparrow_c) + P(\uparrow_c, \uparrow_b). \tag{4.21}$$

Equation (4.21) is called the Bell inequality.

Let us consider observations of spins with a quantum state of entangled two particles with $S = 0$. To begin with, we calculate the probability $P(\uparrow_a, \uparrow_b)$ of Eq. (4.21). If the particle 1 has $+$ along **a**, the particle 2 has necessarily $-$ along **a**. The state $|\downarrow_b\rangle$ of direction **b** is transformed into the state $|\downarrow_a\rangle$ in direction **a** by a rotation around $y$ axis by angle $-\phi$. Using Eq. (1.110), we obtain

$$\begin{aligned} |\downarrow_a\rangle_2 &= D_y^s(-\phi)|\downarrow_b\rangle_2 \\ &= \left\{\cos\left(-\frac{\phi}{2}\right) - i\sin\left(-\frac{\phi}{2}\right)\begin{pmatrix} 0 & -i \\ i & 0 \end{pmatrix}\right\}\begin{pmatrix} 0 \\ 1 \end{pmatrix}_2 \\ &= \left\{\cos\left(\frac{\phi}{2}\right)|\downarrow_b\rangle_2 + \sin\left(\frac{\phi}{2}\right)|\uparrow_b\rangle_2\right\}. \end{aligned} \tag{4.22}$$

Therefore the probability of having $+$ in direction **b** is expressed as

$$|\langle\uparrow_b|\downarrow_a\rangle_2|^2 = \sin^2\left(\frac{\phi}{2}\right). \tag{4.23}$$

The probability of having positive correlation $++$ in observations of the directions **a** and **b** is given by

$$P(\uparrow_a, \uparrow_b) = \sin^2\left(\frac{\phi}{2}\right) \tag{4.24}$$

as a function of the angle $\phi$ between the vectors **a** and **b**. Similarly, the probabilities of positive correlation ++ are

$$P(\uparrow_a, \uparrow_c) = \sin^2\left(\frac{\theta}{2}\right) \tag{4.25}$$

$$P(\uparrow_c, \uparrow_b) = \sin^2\left(\frac{(\phi - \theta)}{2}\right) \tag{4.26}$$

in other combinations of the detectors. The Bell inequality requests

$$\begin{aligned} C(\phi, \theta) &\equiv P(\uparrow_a, \uparrow_c) + P(\uparrow_c, \uparrow_b) - P(\uparrow_a, \uparrow_b) \\ &= \sin^2\left(\frac{\theta}{2}\right) + \sin^2\left(\frac{(\phi - \theta)}{2}\right) - \sin^2\left(\frac{\phi}{2}\right) \end{aligned} \tag{4.27}$$

to be always positive. For simplicity, if we put $\phi = 2\theta$, Eq. (4.27) becomes

$$C(\phi = 2\theta, \theta) = (-)2\sin^2\left(\frac{\theta}{2}\right)\cos\theta. \tag{4.28}$$

It is clear that Eq. (4.28) is negative if $0 < \theta < \frac{\pi}{2}$. Thus the Bell inequality breaks down in quantum mechanics. The most remarkable breakdown occurs for an angle $\theta = \frac{\pi}{3}$ with $C\left(\theta = \frac{\pi}{3}\right) = -0.25$.

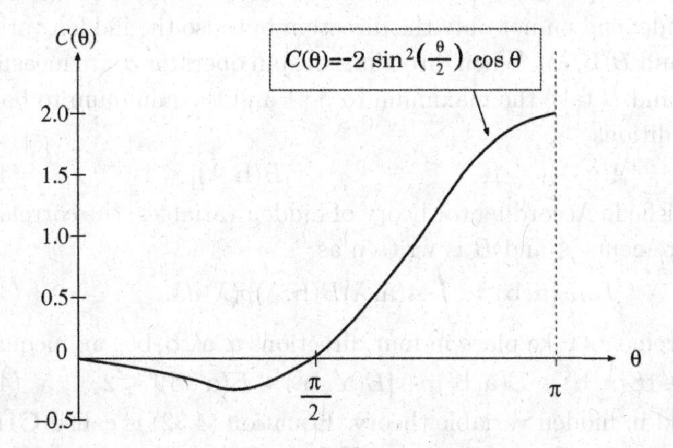


Fig. 4.5 Breakdown of Bell inequality by quantum measurements. The Bell inequality is broken, $C(\theta) < 0$, in an interval of $0 < \theta < \frac{\pi}{2}$.

## 4.4 Measurement of two correlated particles and hidden variable theorem

### 4.4.1 *CHSH inequality*

Theory of hidden variables in quantum mechanics has an idea contrary to statistical interpretation of measurement, and assumes existence of "hidden variables" that determine physical phenomena implicitly. For instance, in measurement of $z$ component of spin in Section 4.3, $x$ component of spin is assigned as a hidden variable. This theory is based upon principle of locality which states, in a system with two particles, the remote action (action in distance) never exists in measurement, i.e., measurement of a particle never affects that of another particle. Instead, it assumes that each particle has its own hidden variable $\lambda$ that have definite values no matter how it is measured or not. The probability of value $\lambda$ is given by a probability density $\rho(\lambda)\quad(\geq 0)$ satisfying the normalization condition:

$$\int \rho(\lambda)\, d\lambda = 1. \tag{4.29}$$

The function $\rho(\lambda)$ is related to the weight function $w(\lambda)$ in (4.10) by $\rho(\lambda) = |w(\lambda)|^2$.

With hidden variable theory, we consider the correlation of spin measurements of EPR pair (4.1) at different positions. As shown by Fig. 4.6, Alice and Bob make measurements with detectors along directions of two vectors $\mathbf{a}$ and $\mathbf{b}$, respectively. Let $A$ and $B$ be results of the measurements. These results depend on not only the direction but also the hidden variable $\lambda$ as $A(\mathbf{a}, \lambda)$ and $B(\mathbf{b}, \lambda)$. When directions of spin operator $\boldsymbol{\sigma}$ are measured, the values $A$ and $B$ take the maximum to be 1 and the minimum to be $-1$. Thus the conditions

$$|A(\mathbf{a}, \lambda)| \leq 1, \qquad |B(\mathbf{b}, \lambda)| \leq 1 \tag{4.30}$$

should be satisfied. According to theory of hidden variables, the correlation of the measurements $A$ and $B$ is written as

$$E_{HV}(\mathbf{a}, \mathbf{b}) = \int A(\mathbf{a}, \lambda) B(\mathbf{b}, \lambda) \rho(\lambda)\, d\lambda. \tag{4.31}$$

When measurements take place in four directions $\mathbf{a}, \mathbf{a}', \mathbf{b}, \mathbf{b}'$ , an inequality

$$S_{HV} \equiv |E(\mathbf{a}, \mathbf{b}) - E(\mathbf{a}, \mathbf{b}')| + |E(\mathbf{a}', \mathbf{b}') + E(\mathbf{a}', \mathbf{b})| \leq 2 \tag{4.32}$$

should be hold in hidden variable theory. Equation (4.32) is called **CHSH inequality** (Clauser, Horne, Shimony, Holt inequality),[1] and can be considered as a variant of the Bell inequality.

[1] J. F. Clauser, M. A. Horne, A. Shimony and R. A. Holt, Phys. Rev. Lett. 23, 880 (1969); J. F. Clauser and M. A. Horne, Phys. Rev. D10, 526 (1974)

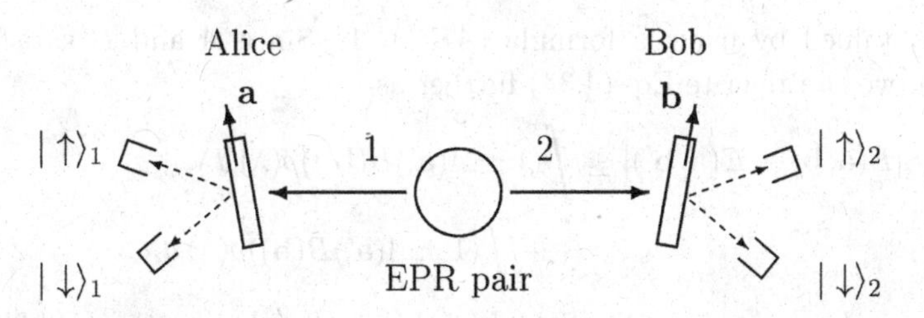


Fig. 4.6 Experiment with EPR pair. Spin projections of two spin $\frac{1}{2}$ particles coupled to total spin $S = 0$ are measured by Alice and Bob measure in directions **a** and **b**.

---

**Example 4.2** Prove CHSH inequality.

**Solution** Hereafter we discard the index of hidden variable $\lambda$, for simplicity, and use symbols $A(\mathbf{a})$, $B(\mathbf{b})$. Firstly, we add and subtract the integral of four variables $A(\mathbf{a}), B(\mathbf{b}), A(\mathbf{a}')$ and $B(\mathbf{b}')$ to the first term on the right-hand side of (4.32);

$$\begin{aligned}|E(\mathbf{a},\mathbf{b}) - E(\mathbf{a},\mathbf{b}')| \\ = \bigg| \int A(\mathbf{a})B(\mathbf{b})\rho(\lambda)\,d\lambda - \int A(\mathbf{a})B(\mathbf{b}')\rho(\lambda)\,d\lambda \\ \pm \bigg[\int A(\mathbf{a})B(\mathbf{b})A(\mathbf{a}')B(\mathbf{b}')\rho(\lambda)\,d\lambda \\ - \int A(\mathbf{a})B(\mathbf{b}')A(\mathbf{a}')B(\mathbf{b})\rho(\lambda)\,d\lambda\bigg]\bigg|. \qquad (4.33)\end{aligned}$$

Note that the sum of the two integrals is zero. Next we rearrange the four integrals as follows;

$$\begin{aligned}|E(\mathbf{a},\mathbf{b}) - E(\mathbf{a},\mathbf{b}')| \\ = \bigg| \int A(\mathbf{a})B(\mathbf{b})(1 \pm A(\mathbf{a}')B(\mathbf{b}'))\rho(\lambda)\,d\lambda \\ - \int A(\mathbf{a})B(\mathbf{b}')(1 \pm A(\mathbf{a}')B(\mathbf{b}))\rho(\lambda)\,d\lambda\bigg| \\ \leq \int |A(\mathbf{a})B(\mathbf{b})|(1 \pm A(\mathbf{a}')B(\mathbf{b}'))\rho(\lambda)\,d\lambda \\ + \int |A(\mathbf{a})B(\mathbf{b}')|(1 \pm A(\mathbf{a}')B(\mathbf{b}))\rho(\lambda)\,d\lambda. \qquad (4.34)\end{aligned}$$

In deriving (4.34), we use an inequality $|a - b| \leq |a| + |b|$ and a relation:

$$|1 \pm AB| = 1 \pm AB \qquad (4.35)$$

which is obtained by using a formula $|AB| \leq 1$. Since $A$ and $B$ satisfy Eq. (4.30), we can rewrite Eq. (4.34) further as

$$\begin{aligned}|E(\mathbf{a},\mathbf{b}) - E(\mathbf{a},\mathbf{b}')| &\leq \int (1 \pm A(\mathbf{a}')B(\mathbf{b}'))\rho(\lambda)\,d\lambda \\ &\quad + \int (1 \pm A(\mathbf{a}')B(\mathbf{b}))\rho(\lambda)\,d\lambda \\ &= 2 \pm (E(\mathbf{a}',\mathbf{b}') + E(\mathbf{a}',\mathbf{b}))\end{aligned} \tag{4.36}$$

using $|AB| \leq 1$. Finally, we can obtain CHSH inequality (4.32) by transferring the second term on the right-hand side of (4.36) to the left-hand side,

$$\begin{aligned}&|E(\mathbf{a},\mathbf{b}) - E(\mathbf{a},\mathbf{b}')| \pm (E(\mathbf{a}',\mathbf{b}') + E(\mathbf{a}',\mathbf{b})) \\ &\leq |E(\mathbf{a},\mathbf{b}) - E(\mathbf{a},\mathbf{b}')| + |E(\mathbf{a}',\mathbf{b}') + E(\mathbf{a}',\mathbf{b})| \\ &\leq 2.\end{aligned} \tag{4.37}$$

Note that the above derivation of CHSH inequality does not involve any probabilistic interpretation, any physical concept about two-particle correlation, nor any principle of quantum mechanics at all.

---

We consider now quantum mechanical measurement of spins in a two-particle system. Suppose that Alice and Bob measure the projection of spin of an EPR pair

$$|\Psi\rangle_{12} = \frac{1}{\sqrt{2}}\{|\uparrow\rangle_1|\downarrow\rangle_2 - |\downarrow\rangle_1|\uparrow\rangle_2\} \tag{4.38}$$

on vectors $\mathbf{a}$ and $\mathbf{b}$. Let us denote the operators for measurements of spins along $z-$ axis by,

$$A(0) = \begin{pmatrix} 1 & 0 \\ 0 & -1 \end{pmatrix}, \qquad B(0) = \begin{pmatrix} 1 & 0 \\ 0 & -1 \end{pmatrix}, \tag{4.39}$$

and

$$A(0)|a\rangle = \pm|a\rangle, \qquad B(0)|b\rangle = \pm|b\rangle, \tag{4.40}$$

where $|a\rangle$ and $|b\rangle$ are eigenstates of $A(0)$ and $B(0)$, respectively, and expressed by the column vectors Eq. (3.15). Let us rotate the detector directions $\mathbf{a}$ and $\mathbf{b}$ around $y-$axis by angles $\theta$ and $\phi$ from the $z-$axis, respectively. The projection of EPR state along declined $\mathbf{a}$ and $\mathbf{b}$ directions can be expressed by

$$|\Psi'\rangle_{12} = D_y^{s1}(\theta)D_y^{s2}(\phi)|\Psi\rangle_{12} \tag{4.41}$$

where the rotation operator $D_y^s(\theta)$ is given by Eq. (3.78);

$$D_y^s(\theta) = \cos\left(\frac{\theta}{2}\right)\mathbf{1} - i\sin\left(\frac{\theta}{2}\right)\sigma_y = \begin{pmatrix} \cos\frac{\theta}{2} & -\sin\frac{\theta}{2} \\ \sin\frac{\theta}{2} & \cos\frac{\theta}{2} \end{pmatrix}. \tag{4.42}$$

It can be easily shown that a single measurement of $A(0)$ or $B(0)$ along **a** or **b** direction is

$$\begin{aligned} \langle\Psi'|A_1(0)|\Psi'\rangle_{12} &= \langle\Psi|D_y^{s2\dagger}(\phi)D_y^{s1\dagger}(\theta)A_1(0)D_y^{s1}(\theta)D_y^{s2}(\phi)|\Psi\rangle_{12} \\ &= \langle\Psi|D_y^{s1\dagger}(\theta)A_1(0)D_y^{s1}(\theta)|\Psi\rangle_{12} \equiv \langle\Psi|A_1(\theta)|\Psi\rangle_{12} \\ &= \langle\Psi|\begin{pmatrix} \cos\theta & \sin\theta \\ -\sin\theta & \cos\theta \end{pmatrix}_1 |\Psi\rangle_{12} = \frac{1}{2}(\cos\theta - \cos\theta) = 0, \end{aligned} \tag{4.43}$$

where $A(\theta)$ is defined by

$$A(\theta) = D_y^{s\dagger}(\theta)A(0)D_y^s(\theta) = \begin{pmatrix} \cos\theta & \sin\theta \\ -\sin\theta & \cos\theta \end{pmatrix}. \tag{4.44}$$

The measurement of EPR pair along the **b** direction by the operator

$$B(\phi) = D_y^{s\dagger}(\phi)B(0)D_y^s(\phi) = \begin{pmatrix} \cos\phi & \sin\phi \\ -\sin\phi & \cos\phi \end{pmatrix} \tag{4.45}$$

gives also zero value for the expectation value. Next, let us evaluate a correlation measurement along **a** and **b** directions. The expectation value of the measurement $E_{QM}(\mathbf{a},\mathbf{b})$ is given by

$$\begin{aligned} E_{QM}(\mathbf{a},\mathbf{b}) &= \langle\Psi'|A_1(0)B_2(0)|\Psi'\rangle_{12} = \langle\Psi|A_1(\theta)B_2(\phi)|\Psi'\rangle_{12} \\ &= -(\cos\theta\cos\phi - \sin\theta\sin\phi) = -\cos(\theta-\phi) = -\mathbf{a}\cdot\mathbf{b} \end{aligned} \tag{4.46}$$

for measurements of spins along **a** and **b** directions.

Suppose that Alice has detectors along the directions **a** and **a**$'$, while Bob has those along **b** and **b**$'$. Place **a** and **a**$'$ with angles $\theta$ and $\theta'$, and **b** and **b**$'$ with $\phi$ and $\phi'$ from $z-$axis. as shown in Fig. 4.7. For CHSH inequality, quantum mechanical observation gives

$$\begin{aligned} S_{QM} &= |E_{QM}(\mathbf{a},\mathbf{b}) - E_{QM}(\mathbf{a},\mathbf{b}')| + |E_{QM}(\mathbf{a}',\mathbf{b}') + E_{QM}(\mathbf{a}',\mathbf{b})| \\ &= |\cos(\theta-\phi) - \cos(\theta-\phi')| + |\cos(\theta'-\phi') + \cos(\theta'-\phi)|. \end{aligned} \tag{4.47}$$

When four detectors placed along **a**, **b**, **a**$'$ and **b**$'$ is arranged with an equal separate angle of $\frac{\pi}{4}$,i.e., $\theta = \frac{\pi}{4}, \phi = \frac{\pi}{2}, \theta' = \frac{3\pi}{4}, \phi = \pi$, we obtain

$$S_{QM} = \cos\left(\frac{\pi}{4}\right) - \cos\left(\frac{3\pi}{4}\right) + \cos\left(\frac{\pi}{4}\right) + \cos\left(\frac{\pi}{4}\right) = 2\sqrt{2} > 2 \tag{4.48}$$

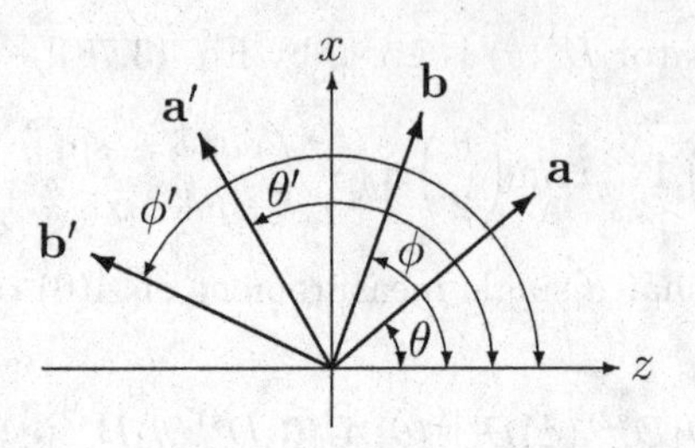


Fig. 4.7 Measurements of spin projections of EPR pair on four directions: $\mathbf{a}, \mathbf{b}, \mathbf{a}', \mathbf{b}'$.

where the CHSH inequality is clearly violated.

The correlation function (4.47) was measured by a hadronic reaction measuring a di-proton pair recently.[2] A proton has a spin 1/2 and the di-proton pair makes a spin-singlet state (4.38) by the strong interaction between two protons. In the measurement, the spin directions of two protons were measured by two polarimeters **a** and **b** with a relative angle $\Phi$. The angle $\Phi$ was varied between $-\pi < \Phi < \pi$ to obtain the correlation function $\mathrm{E}_{exp}$. To compare the results with Bell's inequality, the CHSH correlation function was evaluated to be

$$S_{exp}(\Phi) \equiv |E_{exp}(-\Phi) - E_{exp}(3\Phi)| + |E_{exp}(\Phi) + E_{exp}(\Phi)| \tag{4.49}$$

with $\leq \Phi \leq \pi/2$. Quantum mechanical prediction is $S_{QM} = 3cos(\Phi) - cos(3\Phi)$ for this measurement. In Fig. 4.8, the experimental results are compared with the quantum mechanical prediction and also the upper limit of Bell's inequality. It is very impressive that the experimental data follow the quantal prediction precisely for the entire angle region. The data exceed also Bell' inequality limit $S = 2$ in a wide angle region. The experimental data show the maximum violation of Bell's inequality at the angle $\Phi = \pi/4; S_{exp}(\pi/4) = 2.83 \pm 0.24_{stat} \pm 0.07_{sys}$ with small statistical and systematic errors. As a conclusion of this di-proton experiment, the quantum mechanical strong correlations in the CHSH function are proved in a very high confidence level of 99.3%.

[2]H. Sakai, T. Saito, T. Ikeda, K. Itoh, T. Kawabata, H. Kuboki, Y. Maeda, N. Matsui, C. Rangacharyulu, M. Sasano, Y. Satou, K. Sekiguchi, K. Suda, A. Tamii, T. Uesaka, and K. Yako, Phys. Rev. Lett. **97**, 150405 (2006).

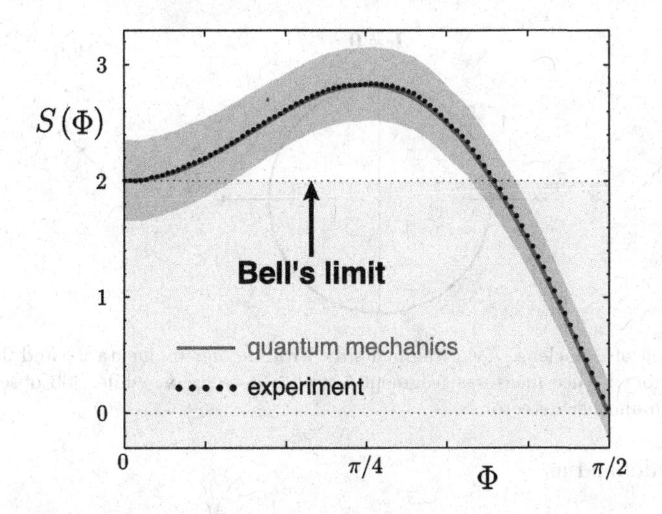


Fig. 4.8 EPR pair and directions of spins measured by Alice and Bob.

### 4.4.2 *Classical correlation vs. quantal correlation: a case of nuclear fission*

Bell inequality is a general theory applicable to study correlations between any two events. It is completely independent of any principles of physics. In other words, it gives an upper bound to the correlation between two distinct events in spacetime as long as locality principle holds. It is a surprising fact that this inequality is broken in quantum mechanics. Let us consider correlation of two events in the case of nuclear fission.

We study in Fig. 4.9 a fission of an atomic nucleus with angular momentum $\mathbf{J} = 0$. A nucleus is broken into two pieces: one with angular momentum $\mathbf{J}_1$ and the other with $\mathbf{J}_2$. The conservation law of angular momentum gives $\mathbf{J} = \mathbf{J}_1 + \mathbf{J}_2 = \mathbf{0}$, which means equal magnitude ($J_1 = J_2$) and opposite signs ($\mathbf{J}_1 = -\mathbf{J}_2$) of two angular momenta. In the case with $J_1 = J_2 = \frac{1}{2}$, we can identify two fragments as an EPR pair.

Alice measures the angular momentum of first fragment with a detector in direction of $\mathbf{a}$, while Bob observes another fragment with a detector of $\mathbf{b}$. We will record measurements of variables $(\mathbf{a} \cdot \mathbf{J}_1)$ and $(\mathbf{b} \cdot \mathbf{J}_2)$ by signs, $+1$ or $-1$ and call them $a$ and $b$, respectively. Namely, we have the results,

$$a \equiv \text{sign}(\mathbf{a} \cdot \mathbf{J}_1) = \text{sign}(\mathbf{a} \cdot \boldsymbol{\sigma}_1) = \pm 1 \tag{4.50}$$

$$b \equiv \text{sign}(\mathbf{b} \cdot \mathbf{J}_2) = \text{sign}(\mathbf{b} \cdot \boldsymbol{\sigma}_2) = \pm 1. \tag{4.51}$$

We repeat this measurement $M$ times, denoting $i$th event as $a_i, b_i$. The

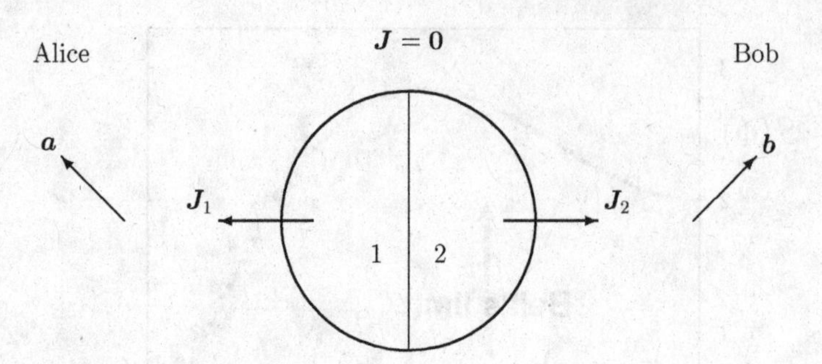


Fig. 4.9 Fission of a nucleus. Two fragments fly with angular momenta $\mathbf{J}_1$ and $\mathbf{J}_2$ to opposite directions. Alice measures fragment 1 with a detector **a**, while Bob observes fragment 2 with another detector **b**.

averages are defined as

$$E_a = \sum_{i=1}^{M} a_i/M, \qquad E_b = \sum_{i=1}^{M} b_i/M. \tag{4.52}$$

Because $\mathbf{J}_1$ and $\mathbf{J}_2$ are both pointed to arbitrary directions, their average values approach 0 as $M$ becomes large enough:

$$E_a \to 0, \qquad E_b \to 0 \qquad (M \to \infty). \tag{4.53}$$

How about the correlation

$$E_{ab} = \sum_{i=1}^{M} a_i b_i/M \tag{4.54}$$

between $a_i$ and $b_i$ ? If the vector **a** is positioned to be parallel to **b**, we have $b_i = -1$ whenever $a_i = 1$ and $b_i = 1$ whenever $a_i = -1$, since $\mathbf{J}_1$ is anti-parallel to $\mathbf{J}_2$. In this case, we get the complete anti-correlation

$$E_{ab} = -1. \tag{4.55}$$

We will derive the correlation when detectors **a** and **b** are placed at angle $\theta$. Let us imagine a sphere where the detector **a** is in the direction of the north pole and the detector **b** inclines by an angle $\theta$ from **a** as shown in Fig. 4.10.

On the sphere with the pole **a**, $a = +1$ if $\mathbf{J}_1$ points to the north hemisphere (shown by a semicircle of a solid line); $a = -1$ of $\mathbf{J}_1$ points to the south hemisphere. On another sphere with the pole of **b**, $b = +1$ if $\mathbf{J}_2$ points to the south hemisphere (shown by a semicircle of a broken line); $b = -1$ if $\mathbf{J}_2$ points to the north hemisphere. Therefore, the sphere is divided by the two equatorial planes into four sectors with alternating signs

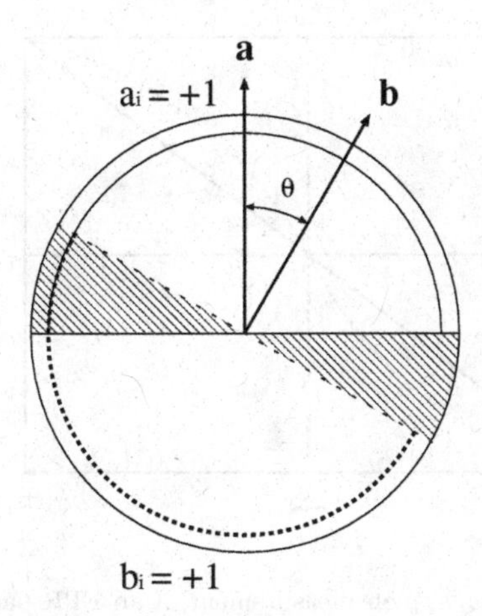


Fig. 4.10 Geometry of correlation of measurements by two detectors **a** and **b**. Vector **a** is placed to the direction of the north pole, while vector **b** inclines by an angle $\theta$ from **a**. This figure shows cross sections divided by equatorials of two spheres whose north poles are assigned by **a** and **b**, respectively. On the sphere with the pole **a**, the value $a = +1$ if $\mathbf{J}_1$ points to the north hemisphere (shown by a semicircle of a solid line); otherwise $a = -1$. On the sphere with the pole **b**, $b = +1$ if $\mathbf{J}_2$ points to the southern hemisphere (shown by a semicircle of a broken line); otherwise $b = -1$. In the shaded area $ab$=1; otherwise $ab = -1$.

of $ab$; the shadowed areas of Fig. 4.10 give $ab = +1$, whereas the other areas correspond to $ab = -1$. Thus we obtain the correlation coefficient

$$E_{ab} = \frac{(2\theta - (2\pi - 2\theta))}{2\pi} = \frac{2\theta}{\pi} - 1. \tag{4.56}$$

It is clear that this result is different from the quantum mechanical correlation (4.46),

$$E_{QM}(\mathbf{a}, \mathbf{b}) = -\mathbf{a} \cdot \mathbf{b} = -\cos\theta \tag{4.57}$$

for projections of spin $\frac{1}{2}$ particles on the directions of detectors **a** and **b** as shown in Fig. 4.11. Results in Fig. 4.11 show that quantum mechanical correlation is always stronger than classical correlation. This result is somewhat surprising since quantum mechanics is believed more "uncertain" than classical mechanics. Namely, classical mechanics is a deterministic theory, while quantum mechanics has intrinsic "uncertainty principle" in observation.

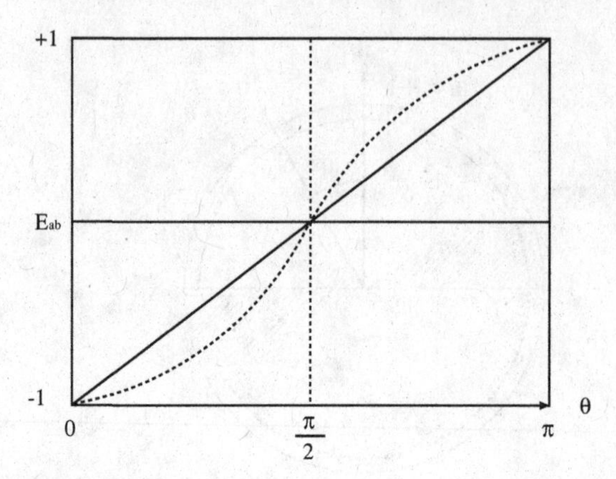


Fig. 4.11 Correlation function $E_{ab}$ on measurement of an EPR pair by two detectors along two vectors $\mathbf{a}, \mathbf{b}$. The solid line shows classical correlation (4.56), while the broken line shows quantum mechanical one (4.57).

## 4.5 EPR experiment by photon pairs

There are many possibilities for EPR experiment. EPR experiment with a pair of spin $\frac{1}{2}$ particles can be done in proton-proton scattering experiments. Recently, another type of experiment is performed by using photon pairs emitted from excited states of atoms. Although the spin of photon is 1, parallel and anti-parallel directions to the direction of motion are only allowed to spin projections of photon because mass of photon is zero. Thus, a pair of photons can make an EPR pair like spin $\frac{1}{2}$ particles since the number of allowed quantum states is 2.

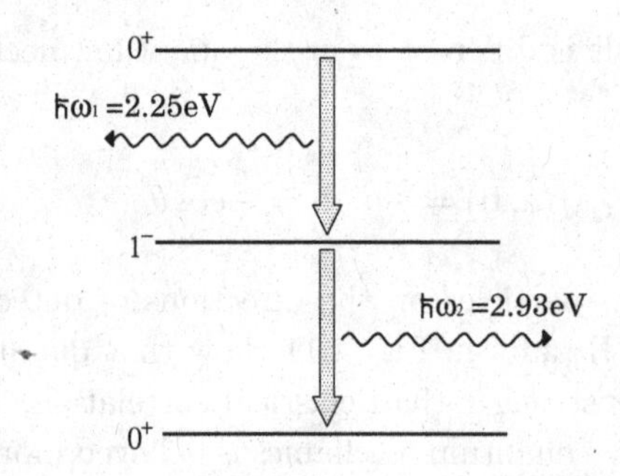


Fig. 4.12 A pair of photons emitted by two consecutive decays of excited states to the ground state of Ca atom: $(J^\pi = 0^+) \to (J^\pi = 1^-) \to (J^\pi = 0^+)$. Two photons have the total spin $S = 0$ and propagate to opposite directions.

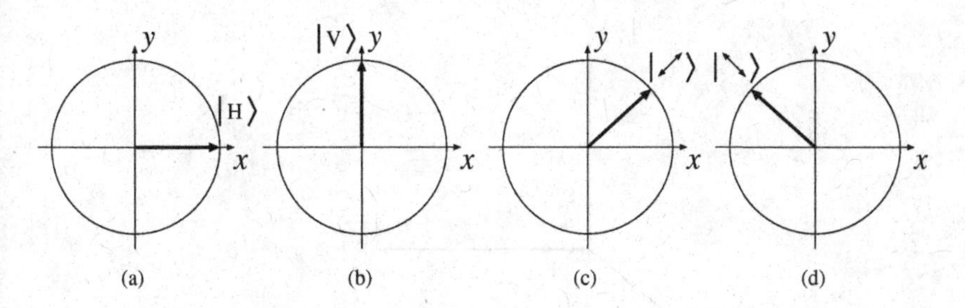


Fig. 4.13 Four polarized states of photons: (a) transverse, (b) longitudinal, (c) diagonal in +45°, (d) diagonal in −45°.

We denote photon polarized along $x$ axis as $|H\rangle$, and that polarized along $y$ axis as $|V\rangle$. These two states are called as linearly polarized states, and expressed by ket vectors as

$$|V\rangle = |\updownarrow\rangle \tag{4.58}$$

$$|H\rangle = |\leftrightarrow\rangle. \tag{4.59}$$

The circularly polarized states are given by

$$|R\rangle = \frac{1}{\sqrt{2}}\{|V\rangle + i|H\rangle\} \text{ :clockwise,} \tag{4.60}$$

and

$$|L\rangle = \frac{1}{\sqrt{2}}\{|V\rangle - i|H\rangle\} \text{ :counterclockwise,} \tag{4.61}$$

by superpositions of the linearly polarized states.

---

**Example 4.3** Show direction of rotation of plane wave propagating in $z$ direction with momentum $k$ and angular velocity $\omega$:

$$\phi^{\pm}(\mathbf{r}, t) = \{\varepsilon_2 \pm i\varepsilon_1\}e^{i(kz-\omega t)} \tag{4.62}$$

where $\varepsilon_1$ and $\varepsilon_2$ are unit vectors of $x$ and $y$ axes, respectively, which stand for linear polarizations as $\varepsilon_1 = |\leftrightarrow\rangle$ and $\varepsilon_2 = |\updownarrow\rangle$.

**Solution** We take the real part of wave function (4.62) in consideration. Then, $x$, $y$ components are given by

$$\phi_x^{\pm}(z, t) = \mp\phi_0 \sin(kz - \omega t) \tag{4.63}$$

$$\phi_y^{\pm}(z, t) = \phi_0 \cos(kz - \omega t). \tag{4.64}$$

Let us look at the wave function from $z$ axis to which the photon is propagating. For a fixed value of $kz$ (for example $kz$=0), $\phi_x^+$ ($\phi_x^-$) becomes positive (negative) from zero value when the clock starts ($t > 0$), while both

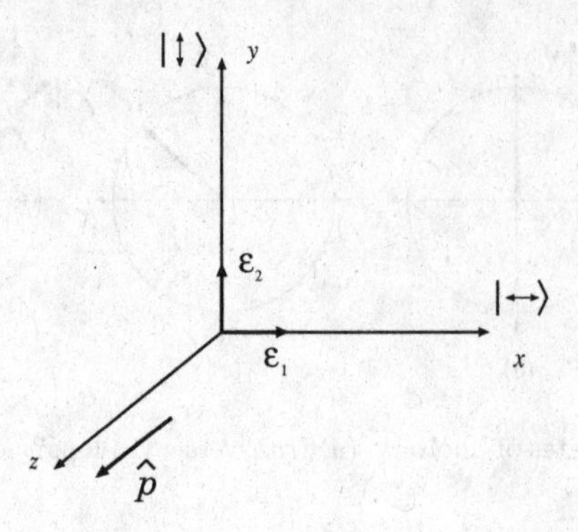


Fig. 4.14 Polarization vectors $\varepsilon_1 and \varepsilon_2$ and momentum vector $\hat{p}$ of photon.

$\phi_y^{\pm}$ decrease from $\phi_0$. This means $\phi^+$ to start rotating clockwise, while $\phi^-$ is rotating counterclockwise. The wave function $\phi^+$ corresponds to right-circularly polarized state $|R\rangle$, while $\phi^-$ corresponds to left-circularly polarized state$|L\rangle$.[3]

---

The unit vector along motion of photon p̂ is given by a vector product:

$$\hat{\mathrm{p}} = \varepsilon_1 \times \varepsilon_2 \tag{4.65}$$

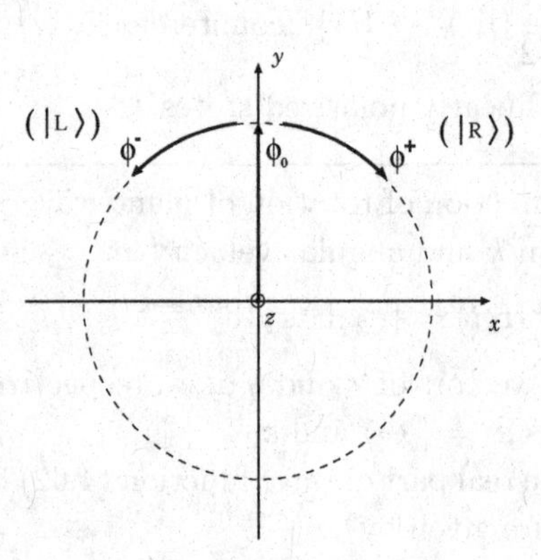


Fig. 4.15 Direction of rotation of circularly polarized state.

[3]In quantum theory of radiation, the state with spin parallel to motion (helicity $h = +1$) is called the right-hand system, while the state with spin anti-parallel to motion (helicity $h = -1$) is called the left-hand system. According to this convention, $\phi^+ = (h = -1)$ is called the left-circularly polarized state, and $\phi^-$ $(h = +1)$ is called the right-circularly polarized state contrary to the convention in optics.

of unit vectors $\varepsilon_1 = |H\rangle$ and $\varepsilon_2 = |V\rangle$. The circularly polarized state is an eigenstate of the **helicity** which is defined by the projection of the spin $\boldsymbol{S}(S=1)$ to the direction of motion of photon $h = (\boldsymbol{S}\cdot\hat{\mathrm{p}})$ whose eigenvalue is $\pm 1$. The counterclockwise polarized state $|L\rangle$ has $h=1$ being parallel to the motion, while the clockwise polarized state $|R\rangle$ has $h=-1$ being anti-parallel to the motion. We consider the consecutive emissions of two photons , $(J^\pi = 0^+) \rightarrow (J^\pi = 1^-) \stackrel{\bullet}{\rightarrow} (J^\pi = 0^+)$ , as shown in Fig. 4.12. Note that the initial state has the angular momentum $J=0$, and the final state also has the angular momentum $J=0$. Therefore the total spin of photon pair must be $S=0$. Momenta of photon EPR pair are opposite in direction so that the helicity must be the same for two photons since spins are also in opposite directions. as shown in Fig. 4.16. Thus, two photons

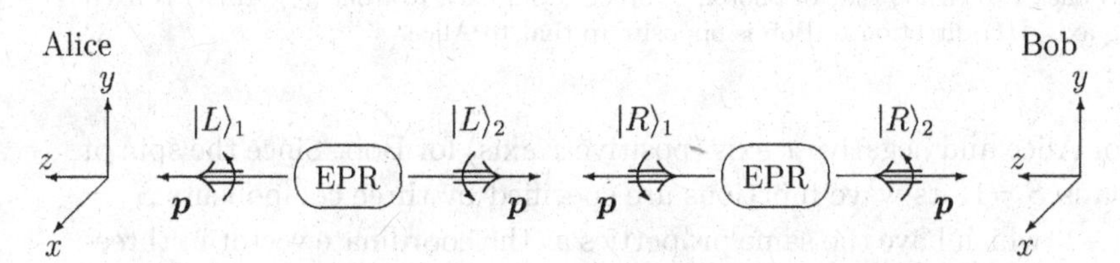


Fig. 4.16 Spins and directions of motion of photon EPR pair. Photon 1 is transmitted to Alice while the photon 2 is transmitted to Bob. In the left circular state $|L\rangle$, the spin is parallel to motion, while the spin is anti-parallel to motion in the right circular state $|R\rangle$.

emitted to opposite directions have the same polarizations $R$ or $L$ and the entangled state is written as

$$|\Psi\rangle_{12} = \frac{1}{\sqrt{2}}\{|R\rangle_1|R\rangle_2 + |L\rangle_1|L\rangle_2\} \tag{4.66}$$

as a superposition of two polarized states. EPR state (4.66) can also be written as

$$|\Psi\rangle_{12} = \frac{1}{\sqrt{2}}\{|H\rangle_1|H\rangle_2 - |V\rangle_1|V\rangle_2\} \tag{4.67}$$

with linearly polarized states. Alice and Bob observe this linearly polarized states as shown in Fig. 4.16. Alice uses a polarimeter arranged at angle $\phi_1$ from the $x$ axis, while Bob uses a polarimeter arranged at $\bar{\phi}_2$ from the $\bar{x}$ axis shown in Fig. 4.17. Since the momentum vector with Alice has the direction opposite to Bob's, the direction $|H\rangle$ ( positive x axis) to Alice is opposite to the direction $|H\rangle$ to Bob. This is because the direction $|H\rangle$ is defined with respect to the intrinsic frame of each photon, i.e., positive $x$

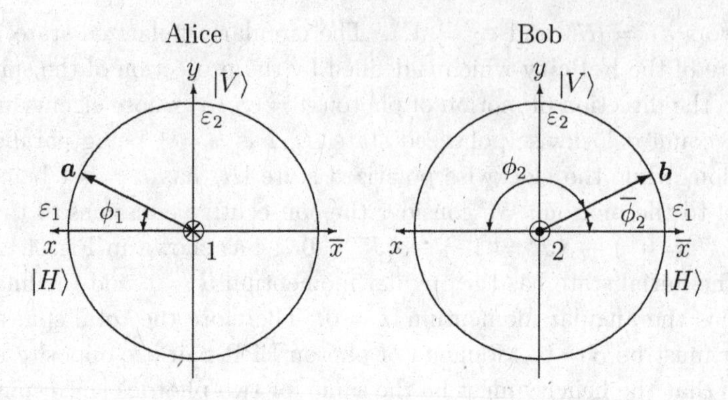


Fig. 4.17 Detectors of Alice and Bob. Momentum vector **p** of photon 1 directs from front to back (⊗), while that of photon 2 directs from back to front (⊙). If $|V\rangle$ is fixed along $y$ axis, $|H\rangle$ direction to Bob is opposite to that to Alice.

axis for Alice and negative $x$ axis (positive $\bar{x}$ axis) for Bob. Since the spin of photon is $S=1$, its wave functions are specified by three components $S_z = +1, 0, -1$, which have the same properties as the coordinate vector in three-dimensional space. We measure the photon polarization by polarimeters **a** and **b** as shown in Fig. 4.17. Let $\phi_1$ be the angle of **a** from $x$ axis, and $\bar{\phi}_2$ be the angle of vector **b** from $\bar{x}$ axis. Let us note that $H$, $V$ and the momentum **p** of photon 1 correspond to $x$, $y$ and $z$ axis of the coordinate system of Alice, while those of photon 2 correspond to $\bar{x}$, $y$ and $z$ axis of the coordinate system of Bob. With the polarimeter **a**, it is evident from Fig. 4.16 that

$$|H_a\rangle = \cos\phi_1|H\rangle_1 + \sin\phi_1|V\rangle_1 \tag{4.68}$$

$$|V_a\rangle = -\sin\phi_1|H\rangle_1 + \cos\phi_1|V\rangle_1. \tag{4.69}$$

Similarly, with the polarimeter **b**, we obtain

$$|H_b\rangle = \cos\bar{\phi}_2|H\rangle_2 + \sin\bar{\phi}_2|V\rangle_2 \tag{4.70}$$

$$|V_b\rangle = -\sin\bar{\phi}_2|H\rangle_2 + \cos\bar{\phi}_2|V\rangle_2. \tag{4.71}$$

Let us evaluate the correlation $E(\mathbf{a}, \mathbf{b})$ in measurements of a photon pair by polarimeters **a** and **b** directed to $\phi_1$ and $\bar{\phi}_2$ respectively. With positive correlations $E_{\leftrightarrow\leftrightarrow}$, $E_{\updownarrow\updownarrow}$ and negative correlations $E_{\leftrightarrow\updownarrow}$, $E_{\updownarrow\leftrightarrow}$ we obtain

$$E(\mathbf{a}, \mathbf{b}) = E_{\leftrightarrow\leftrightarrow} + E_{\updownarrow\updownarrow} - E_{\leftrightarrow\updownarrow} - E_{\updownarrow\leftrightarrow}. \tag{4.72}$$

Each term in Eq. (4.72) is given by

$$E_{\leftrightarrow\leftrightarrow} = \langle\Psi|\{|H_a\rangle_1|H_b\rangle_2\, {}_1\langle H_a|_2\langle H_b|\}|\Psi\rangle_{12}$$
$$= \frac{1}{2}\cos^2(\phi_1 + \bar{\phi}_2), \quad (4.73)$$
$$E_{\updownarrow\updownarrow} = \langle\Psi|\{|V_a\rangle_1|V_b\rangle_2\, {}_1\langle V_a|_2\langle V_b|\}|\Psi\rangle_{12}$$
$$= \frac{1}{2}\cos^2(\phi_1 + \bar{\phi}_2), \quad (4.74)$$
$$E_{\leftrightarrow\updownarrow} = \langle\Psi|\{|H_a\rangle_1|V_b\rangle_2\, {}_1\langle H_a|_2\langle V_b|\}|\Psi\rangle_{12}$$
$$= \frac{1}{2}\sin^2(\phi_1 + \bar{\phi}_2), \quad (4.75)$$
$$E_{\updownarrow\leftrightarrow} = \langle\Psi|\{|V_a\rangle_1|H_b\rangle_2\, {}_1\langle V_a|_2\langle H_b|\}|\Psi\rangle_{12}$$
$$= \frac{1}{2}\sin^2(\phi_1 + \bar{\phi}_2). \quad (4.76)$$

Equation (4.72) is then expressed as

$$E(\mathbf{a},\mathbf{b}) = \cos^2(\phi_1 + \bar{\phi}_2) - \sin^2(\phi_1 + \bar{\phi}_2) = \cos 2(\phi_1 + \bar{\phi}_2). \quad (4.77)$$

With the angle $\phi_2 = \pi - \bar{\phi}_2$, Eq. (4.77) is further rewritten to be

$$E(\mathbf{a},\mathbf{b}) = \cos(2(\phi_1 - \phi_2) - 2\pi) = \cos 2(\phi_1 - \phi_2). \quad (4.78)$$

Comparing with the correlation of the spin $\frac{1}{2}$ particles (4.46), we find that Eq. (4.78) has a positive sign and a factor 2 in front of the angle $\phi_1 - \phi_2$. The positive sign comes from the fact that the polarizations of two EPR photons are the same direction. The factor 2 is explained by the fact that Eq. (4.46) is the correlation of the spin $\frac{1}{2}$ particles, while Eq. (4.78) corresponds to the correlation of photons with spin 1 . Using Eq. (4.78), it is possible to perform the photon experiment on the breakdown of Bell inequality with detector array shown in Fig. 4.7. In particular, the maximum breakdown will be detected with a detector array of $\mathbf{a}$, $\mathbf{b}$, $\mathbf{a}'$, $\mathbf{b}'$ with the equally separate angle $\frac{\pi}{8}$.

## 4.6 Four-photon GHZ states

In recent years, the EPR-pair state of two entangled photons can be easily produced by using laser beam. Moreover, the production of the entangled states with three or four photons, namely, the **Greenburger-Horne-Zeilinger states** (GHZ states) is implemented. The GHZ states, which are considered as more developed forms of the EPR pair, show quantum-mechanical non-locality, and are known to make the **entanglement swapping** which can be used in advanced quantum communication. The GHZ state of four photons is given as a tensor product of two independent EPR pairs as

$$|\phi^i\rangle_{1234} = \frac{1}{\sqrt{2}}\{|H\rangle_1|V\rangle_2 - |V\rangle_1|H\rangle_2\} \otimes \frac{1}{\sqrt{2}}\{|H\rangle_3|V\rangle_4 - |V\rangle_3|H\rangle_4\}, \quad (4.79)$$

where $|H\rangle$ and $|V\rangle$ are single-photon states polarized horizontally and vertically, respectively. As shown in Fig. 4.18, two photons from the two different photon pairs are going toward the **polarized beam splitter (PBS)**. The PBS allows the horizontally polarized light to pass through, while it reflects the vertically polarized light. We set the detectors in the arrangement shown in Fig. 4.18, and perform a coincidence detection of photons $2'$ and $3'$, namely, the two photons are detected only if the both detectors detect photons simultaneously. The result will be either of $|H\rangle'_2|H\rangle'_3$ or $|V\rangle'_2|V\rangle'_3$; i.e., two photons must have a common polarization. To be precise, the state of Eq. (4.79) is projected onto the state:

$$|\phi^f\rangle_{1234} = \frac{1}{\sqrt{2}}\{|H\rangle_1|V\rangle_2|V\rangle_3|H\rangle_4 + |V\rangle_1|H\rangle_2|H\rangle_3|V\rangle_4\} \quad (4.80)$$

by the PBS. This equation shows that if we make a coincidence detection of the states of four photons from the GHZ states using the four detectors in Fig. 4.18, only two states, $HVVH$ and $VHHV$, can be observed out of 16 possible combinations of horizontal ($H$) and vertical ($V$) polarizations. The actual experiment on these four photons in the GHZ state was conducted in 2001.[4] As a result, the states $HVVH$ and $VHHV$ were observed with very high accuracy and their intensities were 200 times stronger than other states.

The observation of the states $HVVH$ and $VHHV$ experimentally confirms the existence of the states with entangled photons. This single experiment, however, does not provide a perfect proof of the four-photon GHZ state, because the state with different a phase:

$$|\psi^f\rangle_{1234} = \frac{1}{\sqrt{2}}\{|H\rangle_1|V\rangle_2|V\rangle_3|H\rangle_4 - |V\rangle_1|H\rangle_2|H\rangle_3|V\rangle_4\} \quad (4.81)$$

also produces the same result of measurement. To distinguish between these two states, we introduce the following diagonally polarized states:

$$|45^+\rangle = \frac{1}{\sqrt{2}}\{|H\rangle + |V\rangle\} \quad (4.82)$$

$$|45^-\rangle = \frac{1}{\sqrt{2}}\{|H\rangle - |V\rangle\}. \quad (4.83)$$

[4] J-W. Pan, M. Daniell, S. Gasparoni, G. Weihs and A. Zeilinger, Phys. Rev. Lett. **86**, 4435 (2001).

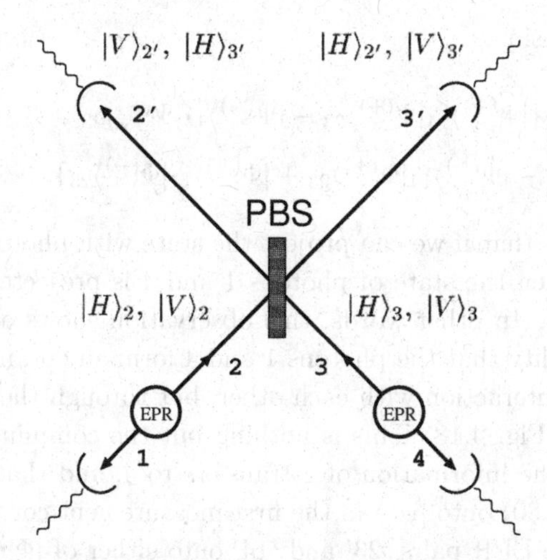


Fig. 4.18 Measurement on the GHZ states by four entangled photons. The photons 2 and 3 from the two different EPR pairs are measured as photons 2′ and 3′ coming through the polarized beam splitter (PBS).

When the state $|\phi^f\rangle$ is expressed in terms of (4.82) and (4.83), the four photons have even number of '+' states. On the other hand, the state $|\psi^f\rangle$, expressed in the same way, has odd number of '+' states. From this argument, we see that the states $|\phi^f\rangle$ and $|\psi^f\rangle$ can be distinguished by the second coincidence measurement with the four detectors in Fig. 4.18 set for the observation of diagonal polarization. In the actual experiment on diagonally polarized states, the coincidence detection of $|++++\rangle$ and $|+++-\rangle$ was made. As a result, it was experimentally confirmed that the former state $|++++\rangle$ was observed in the coincidence measurement with very high probability.

The four-photon GHZ states give rise to an important operation called quantum transfer, or swapping of entangled states, as explained in the following way. By writing the four-photon GHZ state (4.79) in terms of the Bell states analogous to Eqs. (4.7) and (4.8);

$$|\Psi^{(\pm)}\rangle_{12} = \frac{1}{\sqrt{2}}\{|H\rangle_1|V\rangle_2 \pm |V\rangle_1|H\rangle_2\} \tag{4.84}$$

and

$$|\Phi^{(\pm)}\rangle_{12} = \frac{1}{\sqrt{2}}\{|H\rangle_1|H\rangle_2 \pm |V\rangle_1|V\rangle_2\}, \tag{4.85}$$

we obtain the expression:

$$|\psi^i\rangle_{1234} = \frac{1}{2}\{|\Psi^{(+)}\rangle_{14}|\Psi^{(+)}\rangle_{23} - |\Psi^{(-)}\rangle_{14}|\Psi^{(-)}\rangle_{23} - |\Phi^{(+)}\rangle_{14}|\Phi^{(+)}\rangle_{23} + |\Phi^{(-)}\rangle_{14}|\Phi^{(-)}\rangle_{23}\}. \tag{4.86}$$

This equation means that if we can project the state with photons 2 and 3 onto a Bell state, then the state of photons 1 and 4 is projected onto the same Bell state, too. In other words, this observation shows a quantum-mechanical non-locality that the photons 1 and 4 form an entangled states without any direct interaction with each other, but through the photons 2 and 3, as is seen in Fig. 4.18. This is nothing but the communication by swapping in which the information of 2 transfers to 4, and that of 3 to 1. The projection of (4.80) onto $|\psi^f\rangle$ in the first measurement corresponds to the projection of the EPR pairs '23' and '14' onto either of $|\Phi^{(\pm)}\rangle$. In the second measurement, if the states of photons 23 is $++$, the photons 14 is also $++$. This is equivalent to the observation of $|\Phi^{(+)}\rangle$ because the Bell states are transformed as

$$|\Phi^{(+)}\rangle = \frac{1}{\sqrt{2}}\{|++\rangle + |--\rangle\} \tag{4.87}$$

$$|\Phi^{(-)}\rangle = \frac{1}{\sqrt{2}}\{|+-\rangle + |-+\rangle\}. \tag{4.88}$$

In this way, the set of two measurements described above properly confirms the swapping of entangled states by observing one of the Bell states: $|\Phi^{(+)}\rangle$. The observation of four-photon GHZ states was an epoch-making experiment which not only confirmed clearly the non-locality of quantum mechanics , but also opened a door to the new possibility of communication by swapping entangled states.

**Summary of Chapter 4**

(1) Entangled states are realized by photon pairs and di-proton pairs.
(2) EPR pair experiments prove non-locality of observations of entangled states.
(3) Bell's inequality was experimentally proved and justifies strong quantum correlations, which never predicted by neither classical mechanics nor hidden variable theories.
(4) GHZ states open wide possibilities of quantum teleportation.

## 4.7 Problems

**[1]** Prove that the state:

$$|\Psi^{(+)}\rangle = \frac{1}{\sqrt{2}}\{|\uparrow\rangle_1|\downarrow\rangle_2 + |\downarrow\rangle_1|\uparrow\rangle_2\}$$

has the total spin 1.

**[2]** Derive Eq. (4.9).

**[3]** Prove that the correlation (4.56) of fission fragments,

$$E = \frac{2\theta}{\pi} - 1$$

satisfies the CHSH inequality.

# Chapter 5

# Classical Computers

A mathematical model of classical computer was invented by Alan Turing and named "**Turing machine**" model, which is an idealized computer with a simple set of instructions and infinite memories. Soon after Turing's model was proposed, John von Neumann developed a theoretical model for how to implement all the components in a computer to be fully capable as a Turing machine. In more practical way, we will make use of the circuit model, which is useful also in the study of quantum computation. A circuit may involve many inputs, outputs, many wires and many logic gates. These circuits will be implemented by semi-conductors, which have two functions as conductor and insulator and acts under given conditions as a high-speed switch that leads and stops electricity. Modern computers such as personal computers (PC) use integrated circuits (IC) of semiconductor. While the peculiar feature of semiconductor is based entirely on physics of quantum mechanics, we do not call these computers "quantum computers" but rather called "classical computers", because logic gates are based on binary representations and any quantum state of atom or molecule is not used as a logic gate: in classical computers, the data are represented by two logical values 0 and 1 in the circuits using devices with high voltage/low voltage, current on/off, and/or direction of magnetization up/down.

**Keywords: Turing machine; von Neumann computer; computation complexity; logic gate; sequential circuit.**

## 5.1 Turing machine

Algorithms are the most basic concept of computer science. An algorithm is a recipe for performing some task, such as the elementary algebras of adding and subtracting two numbers. A fundamental model for algorithms is Turing machine, which is an ideal machine with a simple set of instructions and an idealized unbound memory. The fundamental question in the study of algorithm is to find algorithms for computational problems , and at the same time providing limitations on the ability to solve them. In practice, there is a large gap between the best model known for solving a problem and the most stringent limitation known for the solution. One may ask a question whether a problem can be solved or not by a computer. Even if the problem is solvable, the next question is whether the answer is obtained in a reasonable length of computational time. Of course the detail depends on ability of specific computers. The **Turing machine** was proposed as a generalized and abstracted mathematical model of the computer. Here we give the basic structure of the Turing machine.

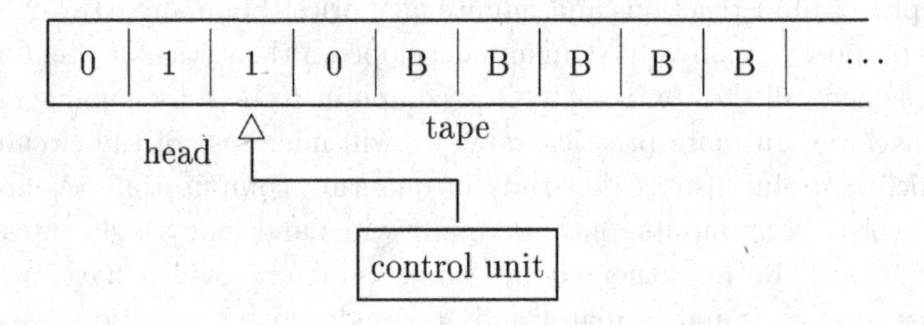


Fig. 5.1 Turing machine.

The structure of the Turing machine is shown in Fig. 5.1. The basic elements of the Turing machine are an infinitely long **tape** and a **control unit** which reads a tape. The tape has cells on which one can write several different kinds of symbols (Fig. 5.1). There are also blank cells (shown by the symbol B). The tape plays a role of a memory device. The control unit has a finite number of states, $0, 1, 2, \cdots, n$, and a **head** which reads and writes symbols on the cell of the tape by moving forth and back (or right and left) along the tape. According to these instructions, the machine will be able to solve a problem by the program written on the tape. In Turing machine, the data are written on the tape and read by the control unit. The tape is the memory device in this machine. The control unit will read and write symbols on the tape and perform computations under the program written on the tape. The data on the tape have definite values and the control unit can read only one datum and perform one instruction on the tape at each process. In this procedure, the Turing machine is deterministic as is the classical mechanics. This is the reason why the computer based on the model of Turing machine is called "classical" computer.

The Turing machine goes on sequentially. The head stops at a certain cell. The symbol written on the same cell and the state of the control unit of the next step are determined as a function of the present symbol and state, together with the direction of the head (right or left). (Fig. 5.2). The

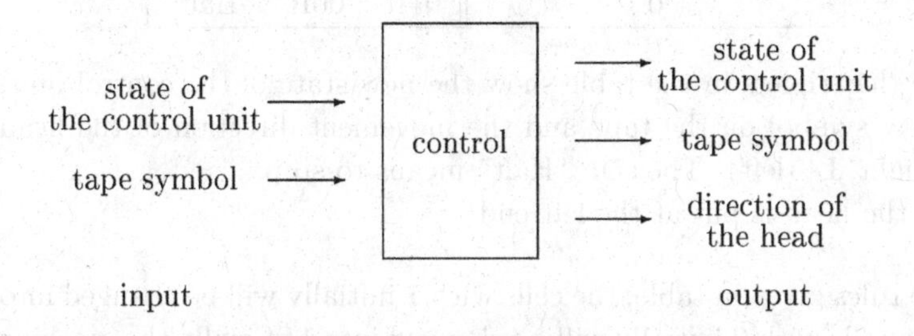


Fig. 5.2 Control of the Turing machine.

control rules are set in advance of running the machine in accordance with what job is assigned to the machine.

The Turing machine operates as follows.

(1) Firstly, the status of the control unit and the position of the head are given.
(2) Symbols are written from the first cell at the left end to a certain cell on the tape in order as the input of the Turing machine.
(3) Once the Turing machine starts operation, it follows the rule and works automatically. In each step, the control unit changes its state, the head writes the symbol on the cell, and the head moves to the right or the left, depending on the state of the control unit and the symbol that the head is reading.
(4) When the control unit and the tape satisfy certain conditions, the machine will stop.
(5) The final symbols on the tape are the output of operation.

---

**Example 5.1** Allow 0 and 1 for the symbols on the cell of the tape. Design a Turing machine which inverts all the symbols on the cells of the tape.

**Solution** For example, the following instructions will work.

- State of control unit: 0 only.
- Symbols on the tape: 0 and 1.
- Operation rules are given in the table:

| state of the control unit | symbol on the tape | | |
|---|---|---|---|
| | 0 | 1 | B |
| 0 | 01R | 00R | Halt |

  where the triplets in the table show the new state of the control unit, the new symbol on the tape and the movement direction of the head (R→right, L→left). The sign "Halt" means to stop.
- First, the head is put at the left end.

By the rules given in table, the cells with 1 initially will be changed into 0 while the cells with 0 initially will be changed into 1. Finally the machine stops by the cell "B".

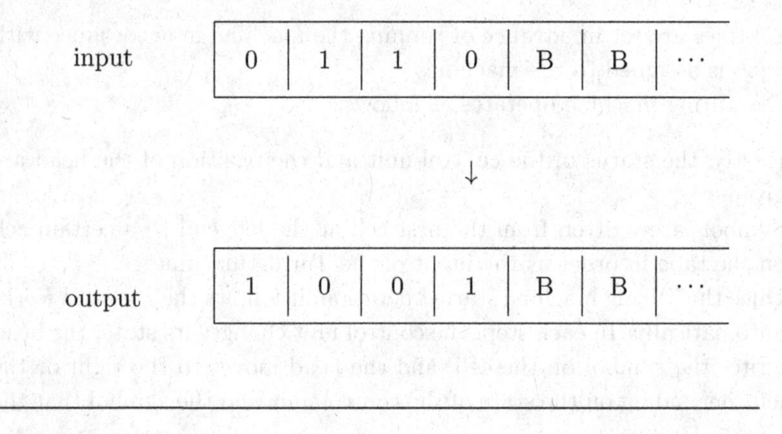


---

**Example 5.2** Design a Turing machine which shifts all the symbols on the tape to the right by one unit. Put 0 to the left end.

**Solution** Assign two states 0 and 1 for the control unit. Every time when the head reads the symbols on cell, write the same symbol on the next cell.

- control unit: 0 and 1
- symbols of the tape: 0 and 1
- rules:

| state of the control unit | symbol on the tape | | |
|---|---|---|---|
| | 0 | 1 | B |
| 0 | 00R | 10R | Halt |
| 1 | 01R | 11R | Halt |

- the initial state is $q = 0$ and the head is at the left end.

In this Turing machine, the control unit acts as a memory.

---

**Example 5.3** Design a Turing machine which shifts symbols on the tape by one unit to the left. Write 0 at the right end. The head is at the left end to start.

**Solution** Since the head is at the left end, the first move will be to the right. Let the control unit have the states given by two bits $(q_0 q_1)$. The bit $q_0$ keeps 0 while the head is moving to the right, and changes into 1 if the head reaches the right end. The bit $q_1$ acts as a memory.

- two bit control unit: $q_i = 0, 1 (i = 0, 1)$
- symbols on the tape: 0 and 1
- rules:

| states of the control unit | symbols on the tape | | |
|---|---|---|---|
| $(q_0 q_1)$ | 0 | 1 | B |
| (00) | (00)0R | (00)1R | (10)BL |
| (01) | (01)0R | (01)1R | (10)BL |
| (10) | (10)0L | (11)0L | Halt |
| (11) | (10)1L | (11)1L | Halt |

- the initial control state is (00) and the head is at the left end.

The head will move from the left end to the right until it reaches the blank (B) cell, at which $q_0$ is changed to 1. Then, the head goes to the left and $q_1$ memorizes the symbol which is read by the head. When the head reaches the left end, the machine stops.

---

We have described one basic type of the Turing machine. The definition of the Turing machine may depend on the literature. Some books give plural numbers of tapes, while some books give a tape without the left end, while the structure is basically the same. The control unit of the Turing machine can be implemented by the logic circuit treated in Section 5.5 (Fig. 5.3).

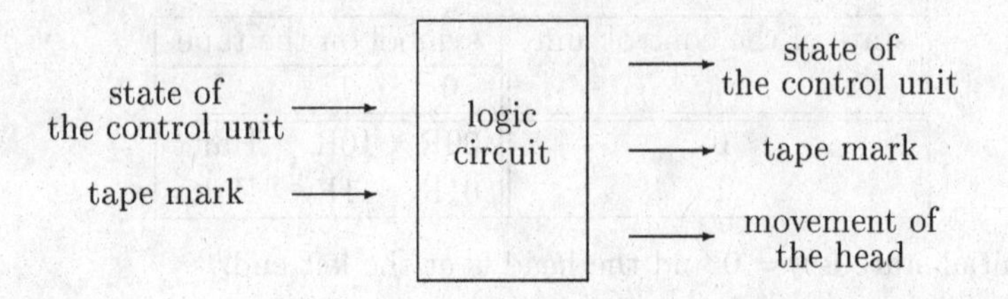


Fig. 5.3 Logic circuit to implement control unit of Turing machine.

So far, we have assumed that the control rules are given in advance. It is possible, however, to give a set of rules in a certain part of the tape. It is called **Universal Turing machine** which is a mathematical model of the present (classical) computer of Neumann architecture.

There is another type of Turing machine which is called **Non-deterministic Turing machine**. In this machine, for the state of control unit and the symbol at the head are given, the next operation is not uniquely determined and there are several choice of the motion in each step. The evolution with non-deterministic Turing machine can be illustrated by a figure with branching.

## 5.2 von Neumann computer

Not long after Turing's proposal for a computer, von Neumann developed a simple model for how to implement all the components of a computer in a practical manner to be capable as a Universal Turing machine. This model is call von Neumann architecture, or von Neumann computer, named after his achievement. All modern computer have the von Neumann architecture. The von Neumann computer has the structure with a memory, a control unit, and input/output devices, as shown in Fig. 5.4. The memory stores a set of instructions (a program) and data. The actual form of the instructions is given as a bit pattern, which governs inputs for gates in a logic circuit of control device. The control device, called **CPU** (Central Process Unit), contains logical and arithmetic circuits. The CPU reads sequentially the instructions in accordance with clock pulses such as Fig. 5.16, controls a flow of data, and performs an operation. If the CPU finds a "jump" instruction, it jumps to a given address instead of reading the next instruction. The set of "instructions" includes "to move a number from the memory to the register," and "to add two numbers in different registers and store the result in another register". Input and output devices include

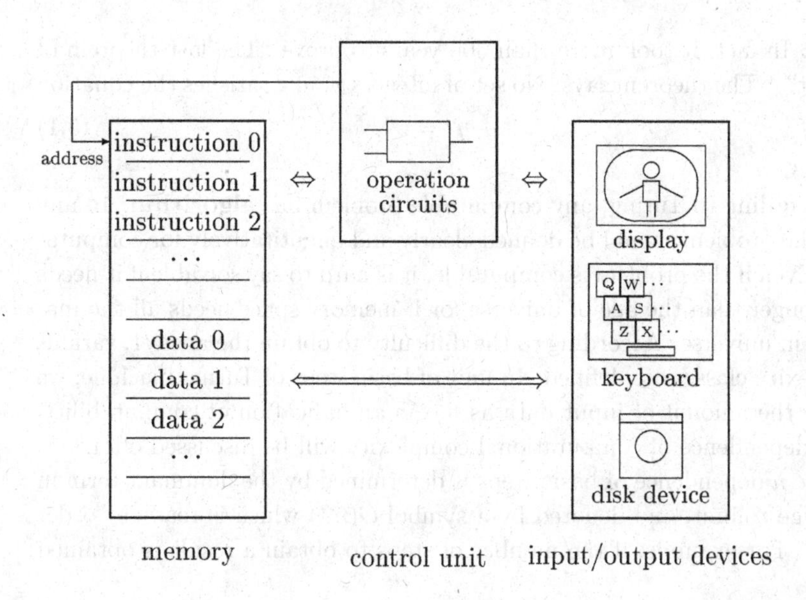


Fig. 5.4 von Neumann architecture.

keyboards, disk drives and displays. This model played a guiding role to realize effective computers in the first stage of history of computer.

## 5.3 Possibility and complexity of computation

What is "computability"? This is a question concerning whether a given problem can be solved by a computer. For example, whether a proposition in mathematics can be proved by a computer? A question to be answered by "yes" or "no" is called **decision problem**: For an element chosen from a set of data, a question is asked whether it belongs to a certain "subset" or not. For example, a question, "Is the number 12 an even number?", is a decision problem. The whole set includes all natural numbers, while a subset consists of all even numbers and another subset has all odd numbers. We may consider the element chosen as an input to the computer, while the output is either 1 (yes) or 0 (no). If a decision problem can be solved, it is called **computable**. If it can not be solved, it is called **incomputable**. In general, it is difficult to prove whether a given problem is computable or not. However it is more difficult to prove that a problem can not be solved. In fact, it took more than 300 years to prove "The last theorem of Fermat".[1] The theorem says: No set of integers $x$, $y$, $z$ satisfies the equation

$$x^n + y^n = z^n \tag{5.1}$$

if $n \geq 3$.

According to Turing, any computable problem has **algorithm**. In any case, the problem should be defined clearly and quantitatively for computation. Even if the problem is computable, it is hard to say solvable if it needs time longer than the age of universe, or if memory space needs all the materials in universe. According to the difficulty to obtain the answer, various complexity classes are defined. In unit of basic steps of Turing machine, we denote the amount of input data as $n$. As an indication of computability, the $n$ dependence of computational complexity will be discussed often.

The $n$ dependence of basic steps is determined by the dominant term in the large $n$ limit, and denoted by a symbol $O(n^\alpha)$ which is read as "order of $n^\alpha$". For example, if the number of steps to obtain a result is obtained as

$$3n(n-1) + 18n \log n,$$

the amount of computation is written as $O(n^2)$, and called "order of $n^2$." This is because the term $n^2$ is dominant in the large $n$ limit and other terms can be discarded. Usually, the coefficient "3"in front of $n^2$ is not important in the large $n$ limit and also neglected. Even when the problem is given, the amount of computation depends on which algorithm is adopted. Therefore, there are many problems about which the amount of computation is not well specified yet.

Now, let us consider what kind of problems can be solved by the computer. The following examples include not only decision problems but also "function problems" for which the outputs can have more than two bits.

### 5.3.1 *Arithmetic operations*

Now, let us consider what kind of problem can be solved by computer. A simple problem is arithmetic operations of integers. A single bit of binary system is equivalent to $\log_{10} 2 \approx 0.3010$ digit of decimal system. In the following, we sometimes take the decimal system which is more familiar to us. Note that there is no fundamental difference between the binary system and the decimal system concerning the amount of computation. The relation between two systems is

[1] The complete proof was shown by Andrew Wiles in 1995.

$$\text{a decimal digit} \approx \text{3.3 binary digits} \equiv \text{3.3 bit information}$$

or

$$\text{a bit information} \equiv \text{a binary digit} \approx \text{0.3 decimal digits} .$$

- Addition and subtraction of integers of $n$ bits
  Let us consider how we perform addition of $n$ bit numbers by hand. We align two numbers vertically and add two numbers at each bit one by one from $2^0$ until the digit of $2^{n-1}$. If there is overflow, we carry one to the next digit. In this way, we can complete the addition in $O(n)$ steps. When there is only one tape in Turing machine, it may take $O(n^2)$ steps because the head takes $O(n)$ steps in commuting between the addend and the augend. If the addend and the augend and the sum are written on separate tapes, the steps of $O(n)$ might be enough to attain the addition. In any case, the addition is attained in the order of polynomial.
- Multiplication and division of $n$-bit integers
  If we consider how we perform multiplication and division manually, the order will be $n^2$ in both cases. About multiplication, some smart algorithms are known to perform it in $O(n \log n \log n \log n)$.

#### 5.3.1.1 *P problem*

Problems like arithmetic operations can be solved within polynomial time of the amount of data $n$, and called **P problem** in the classification of **computational complexity**. A set of linear equations is known as P problem because the number of arithmetic operations to obtain the solutions is bound by a polynomial of the number of unknowns $n$. In general, any problem can be included as P problem when it is solved by steps of any polynomial of data $n$ no matter what the power is.

### 5.3.2 *Factorization and decision of prime numbers*

To find a factor of an integer $N$ of $n$ digits, a straightforward algorithm is to try each integer starting from unity up to $\sqrt{N}$ whether it can divide $N$. If none of them divides $N$, $N$ is identified as a prime number. Thus, for an integer $N \approx 10^n$, the computational complexity of factorization will be $\sqrt{N} \approx 10^{n/2}$ since we need the division of about $10^{n/2}$ times. While the computational time for division depends on the power $n$, the dependence of the whole operation will be not less than $O(10^{n/2}) \approx O(3.16^n)$. To factorize a number of 100 digits, we may find its factors in trials of about $10^{50}$ times if they exist. Even for a supercomputer, it will take time longer than the lifetime of universe. A faster algorithm is known for the factorization based on number theory. We will explain this algorithm in a chapter of quantum computation. Even with the faster algorithm, the order will be $O(\exp(\alpha n^{1/2}(\log_2 n)^{1/2}))$, where $\alpha$ is a positive constant, and the computation is slow enough. To define the amount of computation in a more strict way, the problem is often given in the form of "decision (determination, judgment) problem ( word to be checked)". For example, the problem of factorization will be changed as "For a given integer $N$, judge whether $N$ is a prime number." The algorithm will give the answer "Yes" or "No" and will stop. Even if the answer is "No", finding its factors is another problem.

#### 5.3.2.1 *About the NP problem*

Let us take a problem that needs exponential factorial time or more and for which no algorithm in polynomial time exists. But if a solution of this problem is found, the confirmation is possible in polynomial time. This kind of problem is called **Nondeterministic polynomial time problem**, or **NP problem**. The factorization problem is considered as one of NP problems. If computation time is not bound by polynomial time, the complexity is called exponential time. It does not necessarily mean proportion to the exponential function of $n$, but also the case proportional to the factorial of $n$ is also included in the NP problem. For any problem considered to be exponential time, it may become polynomial if such an invasive algorithm is discovered. In fact, there are many unknown aspects about P or NP problems, including the decision (judgment?) of a prime number.

The decision problem of prime number can be modified into a problem of finding a factor. If a factor is found, it is easy to verify it. The NP problem is such a problem as it is not possible to make the decision, but it is easy to verify the solution within a polynomial time.

### 5.3.3 *Combination problem*

There is a famous problem to find the fastest route to visit a number of houses. To this end, in principle, we need to check all the possible routes. For example, a salesperson visits three houses A, B and C in Fig. 5.5. Let us assume distances between the houses as

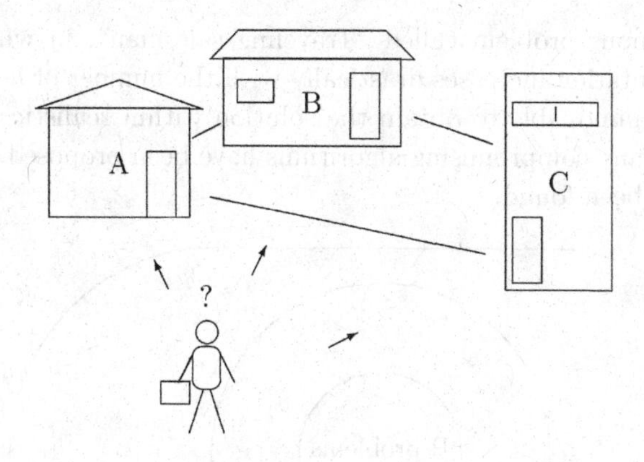


Fig. 5.5 Traveling salesman.

| | distance |
|---|---|
| A to B | 2 km |
| B to C | 3 km |
| C to A | 4 km |

To obtain the shortest route, the straightforward way is to check all the possible routes.

| | route | distance |
|---|---|---|
| (1) | A-B-C | $2 + 3 = 5$ km |
| (2) | A-C-B | $4 + 3 = 7$ km |
| (3) | B-C-A | $3 + 4 = 7$ km |
| (4) | B-A-C | $2 + 4 = 6$ km |
| (5) | C-A-B | $4 + 2 = 6$ km |
| (6) | C-B-A | $3 + 2 = 5$ km |

We thus find that the shortest routes are (1) and (6). Note that the number of routes '6' corresponds to the number of permutation, $3! = 6$.

What will happen if there are 30 houses? In this case, the number of routes is the permutation of 30, i.e., $30! \approx 2.65 \times 10^{32}$. To solve this problem, we have to compute the distances of all the routes. Using a computer with 1 Tela Flops, i.e., $10^{12}$ operations per second, we will need $2.65 \times 10^{20}$ seconds $\approx 8.4 \times 10^{12}$ years to check all the cases. It will be easy to visit 30 houses in a day. However, it is, practically, impossible to find the shortest route because it takes several hundreds times the age of universe ($\approx 10^{10}$ years). We could make shorter the computational time by using 100 computers. A 100 times faster apparatus helps nothing for this problem.

This is a famous problem called "Traveling salesman", in which the amount of computation increases drastically with the number of houses so that it becomes impossible to obtain the solution within realistic computational time. Many compromising algorithms have been proposed, but no decisive one has been found.

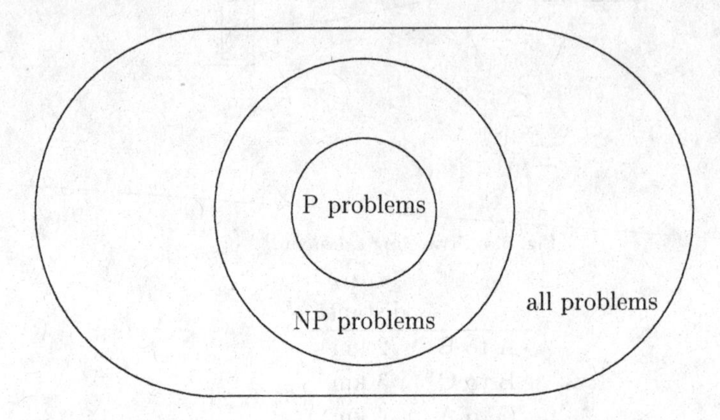


Fig. 5.6 Hierarchy of problems based on computational complexity.

The hierarchy of computational complexity is shown in Fig. 5.6. The P problems, which are actually solvable, are only part of the whole problems. The are many problems outside of NP problems.

#### 5.3.3.1 *NP-complete problems*

Among NP problems, there are so called **NP-complete problems**, which can be converted to other NP-complete problems in polynomial time. It is also known that any NP problem can be attributed to an NP-complete problem. It means that all NP problems are solvable if one of the NP-complete problems is solved. Typical NP-complete problems are

- "Satisfying Problem" (SAT) in Boolean algebra,
- Traveling salesman,
- Hamilton circuit.

The traveling salesman can be treated as a "decision problem" when it is converted to the problem: "Is there a route within a given length?" It is easy to verify if the route is found. It is clear that if a nondeterministic Turing machine can solve the problem in polynomial time, the deterministic Turing machine can solve it. Therefore, the set of P problems is included in the set of NP problems. On the contrary, there is a conjecture that some algorithms to solve in polynomial time can be found for all NP problems. If it is the case, the set of NP problems agrees with the set of P problems.

$$\text{a set of P problems} = \text{a set of NP problems}$$

A question "whether all NP problems are P problems" is known as "P=NP problem", which has not been solved yet.

P problem is commonly considered to be computable. Others are not computable if the size of input is large. Among those incomputable problems for classical computer, quantum computer will be able to solve some of them. In particular, problems to find solutions, which are exponential time problems in classical computer, can be treated as polynomial time problems by a quantum computer, because it can perform simultaneously many trials in parallel processes.

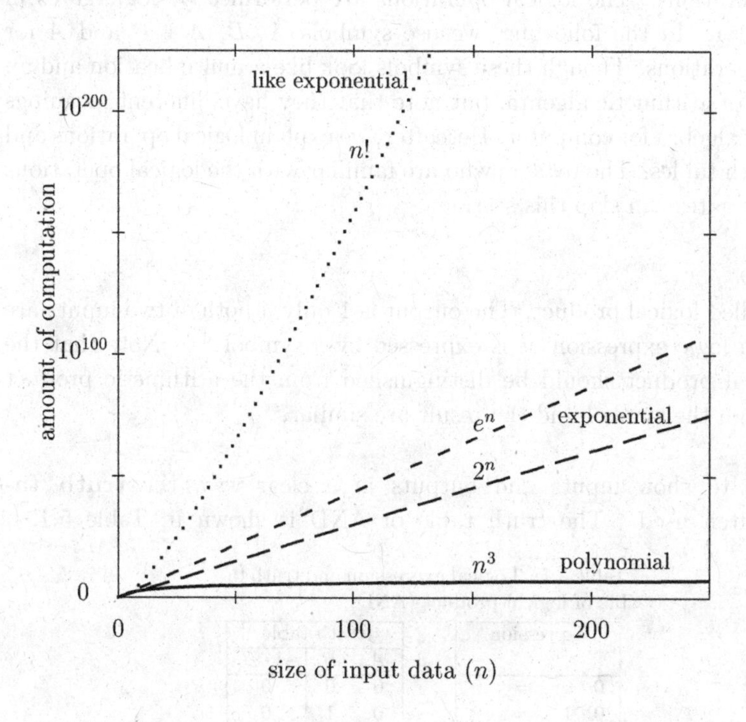


Fig. 5.7 Relation between size of data $n$ and amount of computation.

### 5.3.4 *Computational complexity and amount of computation*

Computational complexity is classified according to the dependence of data size $n$. Algorithm is called **an effective algorithm** if the amount of computation is bound by a polynomial of $n$. Algorithm is called exponential if the amount of computation increases by exponential or more. As seen in Fig. 5.7, the increase of amount of computation is remarkable in the exponential case.

## 5.4 Logic gates

Computers are designed to provide intended output by a combination of logical values 0 and 1. An operation with 0 and 1 is called the logical operation, which is represented by basic operations such as AND, OR and NOT operations. The logical operations are performed by logic gates in a computer. In the following, we use symbols $A \cdot B$, $A + B$ and $\overline{A}$ for logical operations. Though these symbols look like a multiplication and an addition of arithmetic algebra, but note that they have different meanings in logical algebra for computer. Hereafter, we explain logical operations and their truth tables. The readers who are familiar with the logical operations and logic gates can skip this section.

- AND
  is called logical product. The output is 1 only if both of two inputs are 1. In logic expression, it is expressed by a symbol "·". Note that the logical product should be distinguished from the arithmetic product though the symbol and the result are similar.

In order to show inputs and outputs in a clear way, the **truth table** is often used. The truth table of AND is shown in Table 5.1.

Table 5.1 Logical expression and truth table of logical product (AND).

| expression | truth table | | |
|---|---|---|---|
| | $A$ | $B$ | $A \cdot B$ |
| $0 \cdot 0 = 0$ | 0 | 0 | 0 |
| $0 \cdot 1 = 0$ | 0 | 1 | 0 |
| $1 \cdot 0 = 0$ | 1 | 0 | 0 |
| $1 \cdot 1 = 1$ | 1 | 1 | 1 |

**Example 5.4** Show that AND operation satisfies the commutation and the association rules.

**Solution** Table 5.2 shows all possible combinations of values $A$ and $B$ and it is clearly seen that the commutation elation $A \cdot B = B \cdot A$ holds. For the association rule, there are eight combinations and one can see in Table 5.3 that the relation: $(A \cdot B) \cdot C = A \cdot (B \cdot C)$ holds in all cases.

Table 5.2 Commutation rule of AND.

| $A$ | $B$ | $A \cdot B$ | $B \cdot A$ |
|---|---|---|---|
| 0 | 0 | 0 | 0 |
| 0 | 1 | 0 | 0 |
| 1 | 0 | 0 | 0 |
| 1 | 1 | 1 | 1 |

Table 5.3 Association rule of AND.

| $A$ | $B$ | $C$ | $A \cdot B$ | $(A \cdot B) \cdot C$ | $(B \cdot C)$ | $A \cdot (B \cdot C)$ |
|---|---|---|---|---|---|---|
| 0 | 0 | 0 | 0 | 0 | 0 | 0 |
| 0 | 0 | 1 | 0 | 0 | 0 | 0 |
| 0 | 1 | 0 | 0 | 0 | 0 | 0 |
| 0 | 1 | 1 | 0 | 0 | 1 | 0 |
| 1 | 0 | 0 | 0 | 0 | 0 | 0 |
| 1 | 0 | 1 | 0 | 0 | 0 | 0 |
| 1 | 1 | 0 | 1 | 0 | 0 | 0 |
| 1 | 1 | 1 | 1 | 1 | 1 | 1 |

- OR

  is called logical sum. The output is 1 if at least one of the two inputs is 1. The OR operation satisfies the commutation and the association rules. The truth table is shown in 5.4.

Table 5.4 Logical expression and truth table of logical sum (OR).

| expression | | |
|---|---|---|
| $0+0$ | = | 0 |
| $0+1$ | = | 1 |
| $1+0$ | = | 1 |
| $1+1$ | = | 1 |

| $A$ | $B$ | $A+B$ |
|---|---|---|
| 0 | 0 | 0 |
| 0 | 1 | 1 |
| 1 | 0 | 1 |
| 1 | 1 | 1 |

- NOT

Table 5.5 Logical expression and truth table of negation (NOT).

| expression | truth table | |
|---|---|---|
| | $A$ | $\overline{A}$ |
| $\overline{0} = 1$ | 0 | 1 |
| $\overline{1} = 0$ | 1 | 0 |

is called negation. There is one input and one output for this gate and the output is the inverse of input. A symbol $\overline{A}$ means the negation of $A$. The truth table is shown in 5.5.

## 5.5 Logic circuits

Turing machine is an idealized model of computer devices being infinite in size. An alternative model of computation is "**the circuit model**", that is equivalent to the Turing machine , but more realistic and convenient for applications. Especially, the circuit model is important for the study of quantum computers. A circuit is made up of wires and gates, which distribute information around and perform simple computational tasks, respectively. If the bit is passed a "Not" gate, which flips the bit, 1 to 0 or 1 to 0. Table 5.6 gives some examples of circuits.

The logical operation is defined by a combination of elementary operations. **Logic circuits** illustrate operations by circuit symbol as shown in Table 5.6. Each symbol corresponds to a circuit implemented by transistors. A device to implement logical operation is called **logic gate**. For example, the circuit for the operation $C = \overline{A \cdot B}$ is drawn in Fig. 5.8. As

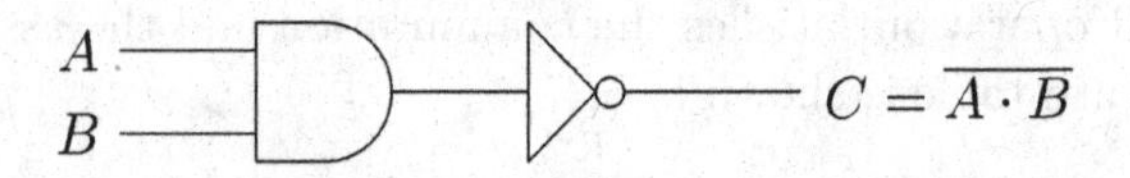


Fig. 5.8 Logic circuit for operation $C = \overline{A \cdot B}$.

another example, we show a circuit called **multiplexer** in Fig. 5.9. The multiplexer selects one input at assigned address from many inputs and delivers it as the output, namely,

* If $A = 0$, the output is $X$;
* If $A = 1$, the output is $Y$.

Table 5.6 Logic circuits performing AND, OR, NOT and XOR gates.

| logical algebra | symbols | logic circuits (input) | logic circuits (output) |
|---|---|---|---|
| AND | $A \cdot B$ | $A$, $B$ | $A \cdot B$ |
| OR | $A + B$ | $A$, $B$ | $A + B$ |
| NOT | $\overline{A}$ | $A$ | $\overline{A}$ |
| XOR | $A \oplus B$ | $A$, $B$ | $A \oplus B$ |

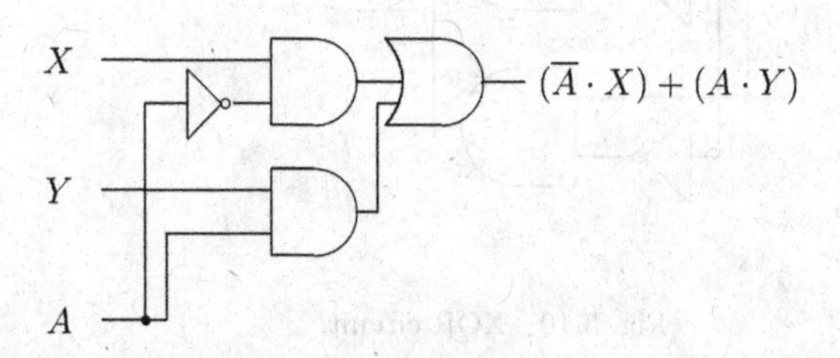


Fig. 5.9 Multiplexer circuit.

This circuit is designed to select one input from two inputs $X$ and $Y$ by the selection given by $A$.

In general, a circuit may involve many inputs and outputs bits

Various gates can be constructed by combinations of the elementary gates explained above. Among them, we give some important ones:

- XOR

is an abbreviation of "Exclusive OR" which is called also "Exclusive logical sum". The symbol is written by $\oplus$, and its logical expressions are given in Table 5.7.

Table 5.7 Logical expressions and truth table of exclusive OR (XOR).

| expression | | | truth table | | |
|---|---|---|---|---|---|
| | | | $A$ | $B$ | $A \oplus B$ |
| $0+0$ | $=$ | $0$ | 0 | 0 | 0 |
| $0+1$ | $=$ | $1$ | 0 | 1 | 1 |
| $1+0$ | $=$ | $1$ | 1 | 0 | 1 |
| $1+1$ | $=$ | $0$ | 1 | 1 | 0 |

The operation XOR is used in the addition of arithmetic algebra, and also in checking parity.

---

**Example 5.5** Design a circuit XOR by combining elementary gates AND, OR and NOT.

**Solution** Figure 5.10 shows an example.

---

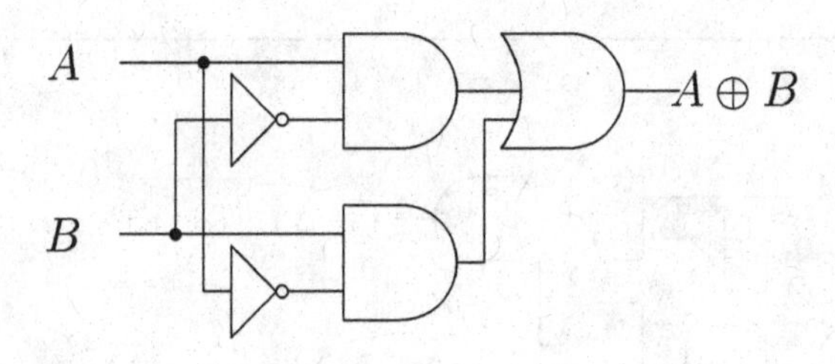


Fig. 5.10 XOR circuit.

As shown by Fig. 5.10, the circuit XOR can be constructed from elementary gates. It has its own circuit symbol as included in Table 5.6.

As an application of the XOR to a circuit, let us take addition circuits (**adders**). The circuit in Fig. 5.11 is called **half adder**. The sum of the two input bits $X$ and $Y$ is given as an output $CS$. The upper bit $C$ is called **carry**. For an addition of numbers with two bits or more, one needs a **full adder** which takes care of carry from the lower digit. The full adder, therefore, has three input bits including the carry. Two output bits $S$ and

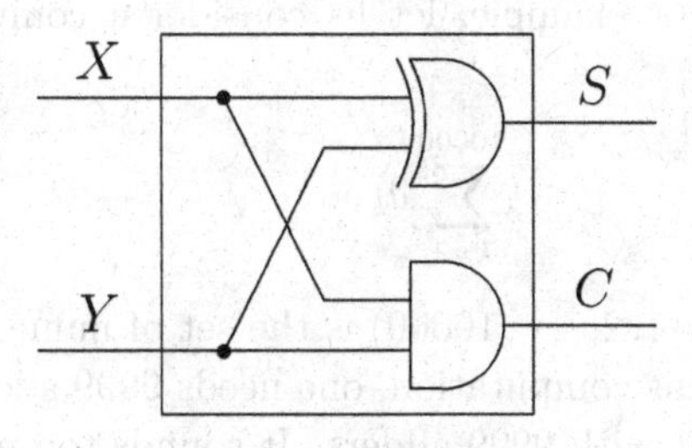


Fig. 5.11 Circuit of half adder.

$C$ are related to three inputs $X$, $Y$ and $Z$ as

$$S = X \oplus Y \oplus Z,$$
$$C = (X \cdot Y) + (Y \cdot Z) + (Z \cdot X),$$

where $S$ is the sum while the other output $C$ is the carry which represents the value delivered to the upper digit as a result of overflow of the original digit.

Let us construct a three-bit adder by combining half adders and full adders, namely, design a circuit for an addition

$$C = A + B,$$

where $A$ and $B$ are the input numbers expressed in three bits:

$$A = A_2 A_1 A_0,$$
$$B = B_2 B_1 B_0$$

and $C$ is the output number in four bits:

$$C = C_3 C_2 C_1 C_0.$$

The designed circuit is shown in Fig. 5.12. Adders for $n$-bit for any number $n$ can be implemented in a similar manner.

## 5.6 Sequential circuit and memory

As shown in Section 5.5, many kinds of circuits to perform logical and mathematical operations can be implemented by combinations of basic gates. We may need, however, tremendous number of gates to construct circuits of various functions. For example, let us consider a computation of the sum:

$$\sum_{k=1}^{10000} a_k, \tag{5.2}$$

where the array $a_k$ ($k = 1, 2, \cdots, 10000$) is the set of numerical data given in advance. To attain this computation, one needs 9999 additions, namely, one would need a circuit with 9999 adders. It sounds too much for such a basic arithmetic addition.

A single adder will be enough, if the addition is performed one by one, just as we do it with a pocket calculator. To do the calculation one by one, we need a device to memorize temporary results. To this end, there are following devices;

- register
  is a device to hold data and commands for a certain period of time.
- memory
  is a device to memorize data.
- counter
  is a device to count the number of pulses; If the logical value of input changes from 0 to 1, the output binary number increases by one. By a special input, one can put intentionally the output number into zero (reset).

This is the essential idea of von Neumann architecture for modern computers. These circuits with memorizing ability are called **sequential circuit**. Let us construct a circuit for summation of Eq. (5.2) with memories. This

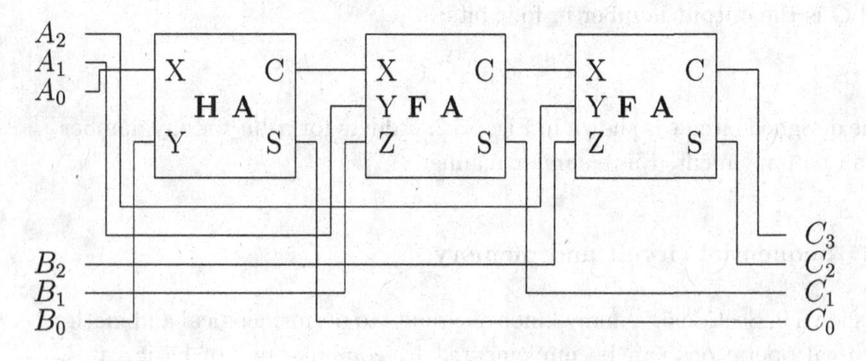


Fig. 5.12 Circuit of three-bit adder (**HA**: half adder; **FA**: full adder).

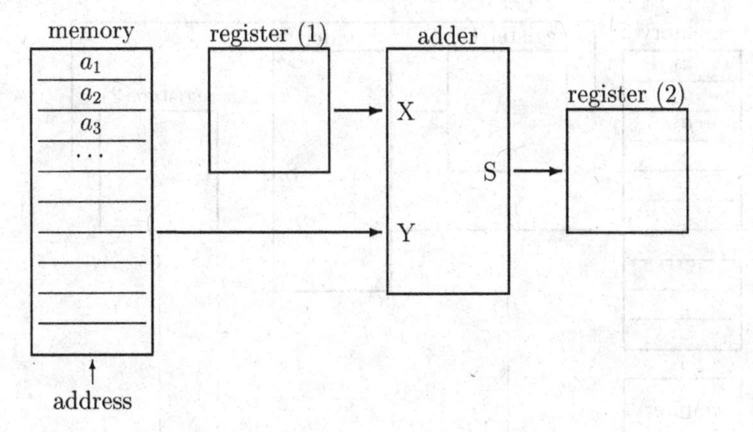


Fig. 5.13 Addition of numbers in memory and register (1). Result is put into register (2).

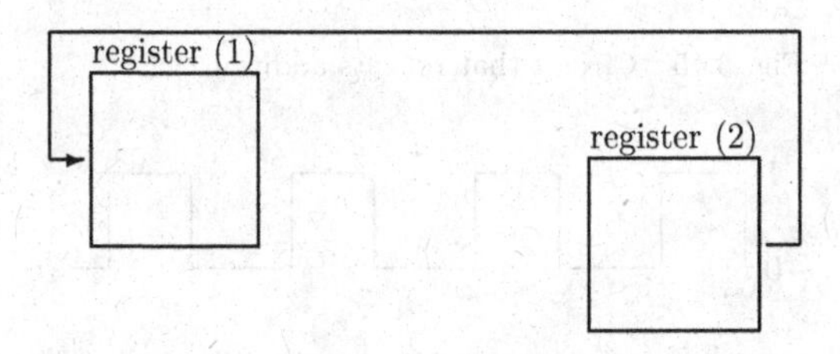


Fig. 5.14 Transfer from register (2) to register (1).

summation can be implemented by repeated action of the following two-step (or two-phase) process.

*(1) Addition*

As shown in Fig. 5.13, one number $Y = a_i$ in a sequential series read from a memory is added to the sum $X = \sum_{k=1}^{i-1} a_k$ stored in register (1). The result $S = X + Y$ is put into register (2).

*(2) Transfer of the result*

As shown in Fig. 5.14, the result in register (2) is transferred to register (1) so that the data in register (1) are replaced by the new result.

The above procedure will be performed automatically in a circuit of Fig. 5.15. The "pulse" means a device that emits logical pulse in time

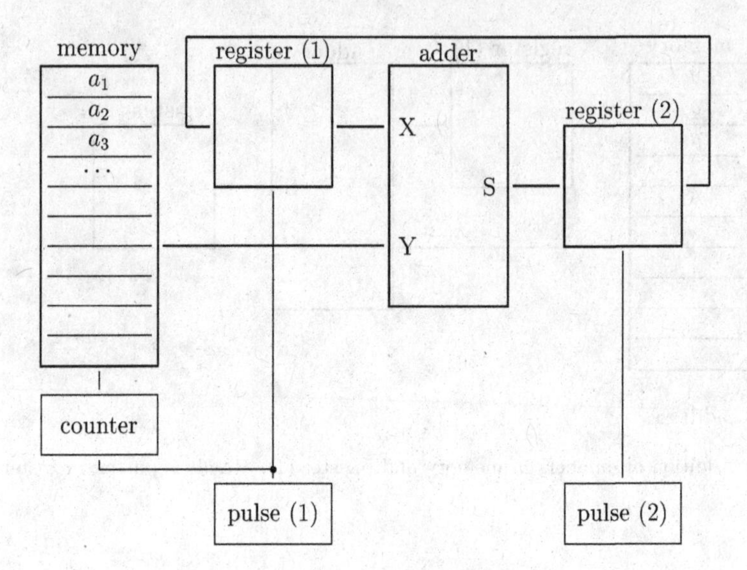


Fig. 5.15 Circuit that repeats addition.

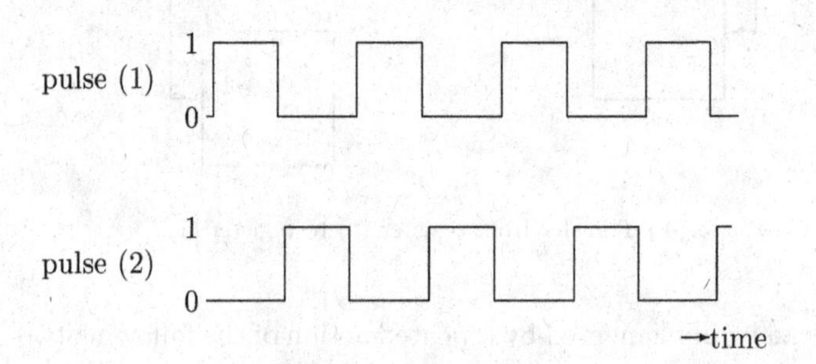


Fig. 5.16 Two kinds of logical pulses.

interval shown in Fig. 5.16. This is called "clock pulse". Each register renews data only when it receives the signal "1" coming from the pulse generator. Two pulse generators give "1" alternately, and they must not give "1" simultaneously.

Computation of the sum (5.2) will be proceeded as follows;

- step(1) Firstly, store numbers $a_i$ in memories, while set register (1) and the counter $i$ equal to zero.
- step(2) Logical pulse starts.
- step(3) If the pulse (1) is 1, the counter $i$ is increased by one unit. Then, read X from register (1) and Y=$a_i$ from memory and obtain S=X+Y .

- step(4) Put S in the register (2) if the pulse (2) is 1.
- step(5) Replace the data in register (1) by the data in register (2).
- step(6) Repeat steps (3)−(5) until the counter exceeds ten thousand and the pulse is stopped.
- step(7) Check the number in register (1) where the sum (5.2) is stored.

The circuit described above is a machine specialized for "addition of ten thousand numbers". Circuit must be arranged in a different way according to what kind of calculation is needed. If one want to do multiplication, one must put out arranged wiring and reconstruct the circuit in a suitable way for multiplication. In fact, some computers of earlier stage used to be organized such a way.

Instead of changing wiring, we can store the wiring diagram in a memory, namely, store the instruction of what should be done in each step. A typical style of the instruction is given together with an address.

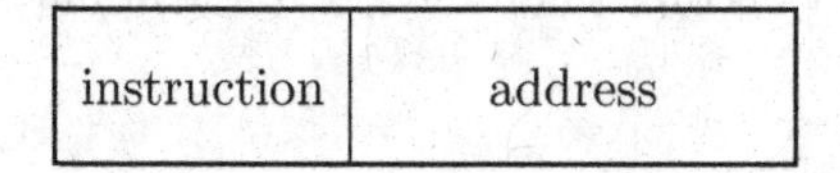


where "instruction" is, for example, "to add a number from the memory", or "to read data in the memory and put it in the register", while "address" gives the location in the memory. The actual form of instruction is a bit pattern. Thus, the operation of computer goes like

read instruction → change of circuit → execution → read instruction → … .

The procedure of computation is given by a program, according to which a computer itself controls the switch inside. A typical machine is called von Neumann computer as was described in Section 5.2.

**Summary of Chapter 5**

(1) Most computers in present use are classified as classical computers.
(2) The Turing machine is an abstract computer model with a set of instructions and a finite length of memory
(3) Most modern computers with stored programs are designed based on the von Neumann architecture.
(4) Basic operations in the computers are implemented by logic gates.
(5) Time-dependent operations of the computers are controlled with sequential circuits.
(6) There are several classification in the complexity of computation. Some kinds of problems are too complex to solve even with the fastest supercomputers today.

## 5.7 Problem

[1] Design Turing machine such that the input is an integer $n$ in binary form and the output on tape is an integer $n+1$ in binary form.

# Chapter 6

# Quantum Gates

As was described in Chapter 3 on "Basic Concepts of Quantum Mechanics", a quantum bit (qubit) might be implemented by two quantum states with two different eigenenergies. The logical values 0 and 1 can be obtained, for example, by the direction of spin of an electron or an atom, the ground state and an excited state of an atom, and also the slit that a photon pass through.

Quantum computer makes use of quantum systems which obey the time-dependent Schrödinger equation. The input state changes into the output state after a period of time governed by the principle of quantum mechanics. Since the Schrödinger equation is invariant under time reversal, the quantum circuit is also time-reversal, i.e., the output state can be reversed to the input states. All gates are time reversible in the quantum computer. In the reversible gate, the number of output terminals must not be smaller than that of input terminals. Otherwise the input can not be recovered. A special character of quantum computer is "the superposition " of logical values 0 and 1. In the classical circuit, like AND gate, the input data is processed to intermediate gates and delivered to the output gates. In the quantum computer, on the other hand, the input and the output are not necessarily distinguished by locations. For example, a spin in a certain input state is affected by an interaction of magnetic field, and change into the output state (Fig. 6.1). Having these characters in mind, we will look at quantum logic circuits. This chapter is devoted to describe the basic circuits which perform quantum computations.

**Keywords: Hadamard gate; controlled-operation gates; quantum Turing machine quantum Fourier transformation.**

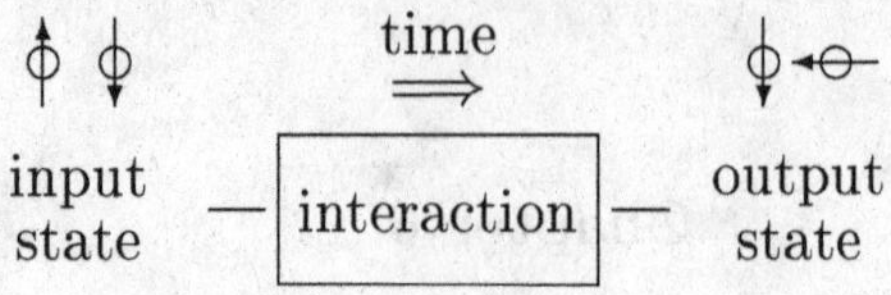

time $\Longrightarrow$

input state — interaction — output state

Fig. 6.1 Action of a quantum gate.

## 6.1 Quantum gates

Qubits are represented by state vectors

$$|0\rangle = \begin{pmatrix} 1 \\ 0 \end{pmatrix}, \tag{6.1}$$

$$|1\rangle = \begin{pmatrix} 0 \\ 1 \end{pmatrix}, \tag{6.2}$$

and a superposition of two states,

$$|\psi\rangle = \alpha|0\rangle + \beta|1\rangle = \begin{pmatrix} \alpha \\ \beta \end{pmatrix} \tag{6.3}$$

where $\alpha$ and $\beta$ are complex numbers. Particularly, quantum gate (6.3) is different from classical gate. Operation is proceeded by changing the state vector by a unitary operator:

$$U \equiv \begin{pmatrix} \langle 0|U|0\rangle & \langle 0|U|1\rangle \\ \langle 1|U|0\rangle & \langle 1|U|1\rangle \end{pmatrix}. \tag{6.4}$$

One can write the action of operator $U$ as

$$\text{output vector} = U \times \text{input vector}, \tag{6.5}$$

as shown in Fig. 6.1. For the input vector (6.3), the output vector is given by

$$\begin{pmatrix} \langle 0|U|0\rangle & \langle 0|U|1\rangle \\ \langle 1|U|0\rangle & \langle 1|U|1\rangle \end{pmatrix} \begin{pmatrix} \alpha \\ \beta \end{pmatrix} = \begin{pmatrix} \langle 0|U|0\rangle\alpha + \langle 0|U|1\rangle\beta \\ \langle 1|U|0\rangle\alpha + \langle 1|U|1\rangle\beta \end{pmatrix}, \tag{6.6}$$

or

$$U|\psi\rangle = (\langle 0|U|0\rangle\alpha + \langle 0|U|1\rangle\beta)\,|0\rangle + (\langle 1|U|0\rangle\alpha + \langle 1|U|1\rangle\rangle\beta)\,|1\rangle.$$

To express the operation of classical gate, we have used truth table. For quantum gate, a matrix representation is also useful. There are following quantum gates;

- identity
  works as $|0\rangle \to |0\rangle$, $|1\rangle \to |1\rangle$. It is represented by the unit matrix:

$$\mathbf{I} = \begin{pmatrix} 1 & 0 \\ 0 & 1 \end{pmatrix}. \tag{6.7}$$

- negation
  converts $|0\rangle$ into $|1\rangle$ and $|1\rangle$ into $|0\rangle$. It is represented by

$$\begin{pmatrix} 0 & 1 \\ 1 & 0 \end{pmatrix}. \tag{6.8}$$

- phase
  puts a phase (see, Fig. 6.2) for a qubit. This gate is unique to qubit.

$$U(\phi) = \begin{pmatrix} 1 & 0 \\ 0 & e^{i\phi} \end{pmatrix} \tag{6.9}$$

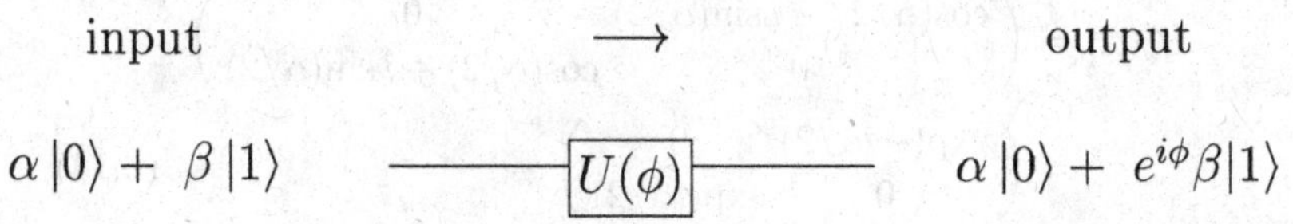

input $\longrightarrow$ output

$\alpha\,|0\rangle + \beta\,|1\rangle$ —— $\boxed{U(\phi)}$ —— $\alpha\,|0\rangle + e^{i\phi}\beta|1\rangle$

Fig. 6.2 Phase gate. Coefficients $\alpha$ and $\beta$ are arbitrary complex numbers.

- Hadamard transformation
  is represented by the matrix:

$$H \equiv \frac{1}{\sqrt{2}} \begin{pmatrix} 1 & 1 \\ 1 & -1 \end{pmatrix}, \tag{6.10}$$

and drawn with $H$ in circuits (see, Fig. 6.3). This is also unique to quantum bits, and one of the circuits that is commonly used. Twice operations of this circuit is equivalent to identity.

- Spin rotation
  is the operator to rotate the spin 1/2 state described in Section 3.5 "Angular momentum, spin and rotation";

$$D_j^s(\alpha) = \cos(\alpha/2) \begin{pmatrix} 1 & 0 \\ 0 & 1 \end{pmatrix} - i\,\sin(\alpha/2)\sigma_j \tag{6.11}$$

| input | $\longrightarrow$ | output |
|---|---|---|
| $\alpha\|0\rangle + \beta\|1\rangle$ | — $H$ — | $\frac{1}{\sqrt{2}}(\alpha+\beta)\|0\rangle$ $+\frac{1}{\sqrt{2}}(\alpha-\beta)\|1\rangle$ |

Fig. 6.3 The Hadamard transformation. Coefficients $\alpha$ and $\beta$ are arbitrary complex numbers.

and is often used in quantum circuits.

The matrix forms of rotations around the $x$-, $y$- and $z$-axis are given as,

$$D_x^s(\alpha) = \begin{pmatrix} \cos(\alpha/2) & -i\sin(\alpha/2) \\ -i\sin(\alpha/2) & \cos(\alpha/2) \end{pmatrix}, \tag{6.12}$$

$$D_y^s(\alpha) = \begin{pmatrix} \cos(\alpha/2) & -\sin(\alpha/2) \\ \sin(\alpha/2) & \cos(\alpha/2) \end{pmatrix}, \tag{6.13}$$

$$\begin{aligned} D_z^s(\alpha) &= \begin{pmatrix} \cos(\alpha/2) - i\sin(\alpha/2) & 0 \\ 0 & \cos(\alpha/2) + i\sin(\alpha/2) \end{pmatrix} \\ &= \begin{pmatrix} \exp(-i\alpha/2) & 0 \\ 0 & \exp(i\alpha/2) \end{pmatrix}. \end{aligned} \tag{6.14}$$

---

**Example 6.1** Show that the following relations hold:

$$D_y^s(\pi/2)\,\sigma_z\,D_y^s(-\pi/2) = \sigma_x \tag{6.15}$$

$$D_y^s(\pi/2)\,\sigma_z = H \tag{6.16}$$

$$D_y^s(-\pi/2)\,\sigma_x = H \tag{6.17}$$

**Solution**

$$\begin{aligned} &D_y^s(\pi/2)\,\sigma_z\,D_y^s(-\pi/2) \\ &= \begin{pmatrix} \cos(\pi/4) & -\sin(\pi/4) \\ \sin(\pi/4) & \cos(\pi/4) \end{pmatrix} \begin{pmatrix} 1 & 0 \\ 0 & -1 \end{pmatrix} \begin{pmatrix} \cos(\pi/4) & \sin(\pi/4) \\ -\sin(\pi/4) & \cos(\pi/4) \end{pmatrix} \\ &= \frac{1}{2}\begin{pmatrix} 1 & -1 \\ 1 & 1 \end{pmatrix} \begin{pmatrix} 1 & 0 \\ 0 & -1 \end{pmatrix} \begin{pmatrix} 1 & 1 \\ -1 & 1 \end{pmatrix} \\ &= \begin{pmatrix} 0 & 1 \\ 1 & 0 \end{pmatrix} = \sigma_x. \end{aligned}$$

This means that the rotation of $\sigma_z$ about $y$-axis by $\pi/2$ makes $\sigma_x$. As for (6.16) and (6.17), we have

$$\begin{aligned} D_y^s(\pi/2)\,\sigma_z &= \begin{pmatrix} \cos(\pi/4) & -\sin(\pi/4) \\ \sin(\pi/4) & \cos(\pi/4) \end{pmatrix} \begin{pmatrix} 1 & 0 \\ 0 & -1 \end{pmatrix} \\ &= \frac{1}{\sqrt{2}} \begin{pmatrix} 1 & -1 \\ 1 & 1 \end{pmatrix} \begin{pmatrix} 1 & 0 \\ 0 & -1 \end{pmatrix} \\ &= \frac{1}{\sqrt{2}} \begin{pmatrix} 1 & 1 \\ 1 & -1 \end{pmatrix} = H, \end{aligned}$$

$$\begin{aligned} D_y^s(-\pi/2)\,\sigma_x &= \begin{pmatrix} \cos(\pi/4) & \sin(\pi/4) \\ -\sin(\pi/4) & \cos(\pi/4) \end{pmatrix} \begin{pmatrix} 0 & 1 \\ 1 & 0 \end{pmatrix} \\ &= \frac{1}{\sqrt{2}} \begin{pmatrix} 1 & 1 \\ -1 & 1 \end{pmatrix} \begin{pmatrix} 0 & 1 \\ 1 & 0 \end{pmatrix} \\ &= \frac{1}{\sqrt{2}} \begin{pmatrix} 1 & 1 \\ 1 & -1 \end{pmatrix} = H. \end{aligned}$$

These equations can also be derived by using commutation and anticommutation relations like $\sigma_i\sigma_j + \sigma_j\sigma_i = 2\delta_{i,j}$ with Eq. (6.11), without writing the matrix elements explicitly.

---

The control gates operate on more than two bits including the control bit and the output bit.

- Controlled-NOT
  inputs $A$, $B \to$ outputs $A'$, $B'$;

$$A' = A \tag{6.18}$$

$$B' = A \oplus B \tag{6.19}$$

In this gate, the output $A'$ is the same as $A$, while $B'$ depends on $A$, namely,

– If $A = 0$, $B' = B$;
– If $A = 1$, $B' = \overline{B}$.

The circuit of controlled-NOT is drawn in Fig. 6.4, and its circuit symbol is shown in Fig. 6.5 together with the truth table.

---

**Example 6.2** E Verify that the controlled-NOT is reversible by expressing the inputs $A$ and $B$ as a function of the outputs $A'$ and $B'$.

**Solution** The output can be obtained by

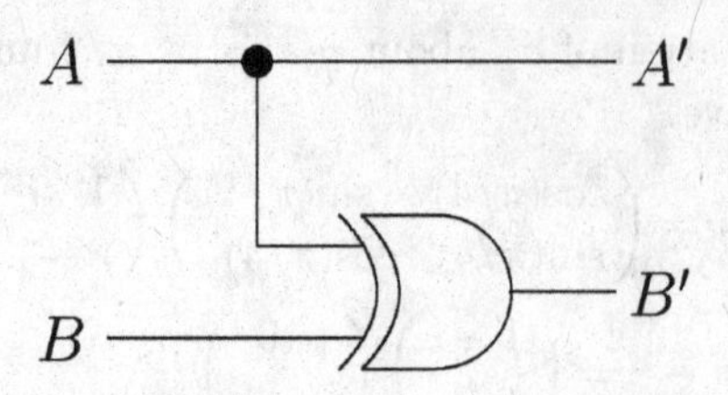


Fig. 6.4 Controlled-NOT.

| $A$ | $B$ | $A'$ | $B'$ |
|---|---|---|---|
| 0 | 0 | 0 | 0 |
| 0 | 1 | 0 | 1 |
| 1 | 0 | 1 | 1 |
| 1 | 1 | 1 | 0 |

The truth table of controlled-NOT

input $\longrightarrow$ output

$A$ — $A'$

$B$ — $B'$

The circuit symbol of controlled-NOT

Fig. 6.5 Truth table and circuit symbol of controlled-NOT.

$$A = A'$$
$$B = A' \oplus B'.$$

The above equations prove that the gate is reversible.

---

In general, multi-qubit states are expressed by **direct products** of vectors. For two-qubit states, for example, the input and the output are expressed by vectors of $2 \times 2 = 4$ elements. Let us write the two qubit state as $|q_A\rangle \otimes |q_B\rangle$, or simply as $|q_A q_B\rangle$, where $|q_A\rangle$ and $|q_B\rangle$ are the eigenstates of operators $A$ and $B$, respectively, and the eigenvalues are either $|0\rangle$ or $|1\rangle$. We express the basis vector as a column vector of four elements and the 2-bit basis states can be expressed as

$$|00\rangle = \begin{pmatrix} 1 \\ 0 \\ 0 \\ 0 \end{pmatrix}, \ |01\rangle = \begin{pmatrix} 0 \\ 1 \\ 0 \\ 0 \end{pmatrix}, \ |10\rangle = \begin{pmatrix} 0 \\ 0 \\ 1 \\ 0 \end{pmatrix}, \ |11\rangle = \begin{pmatrix} 0 \\ 0 \\ 0 \\ 1 \end{pmatrix}. \qquad (6.20)$$

Let us denote the matrix for the controlled-NOT as $U$ Since

$$U|00\rangle = |00\rangle \tag{6.21}$$
$$U|01\rangle = |01\rangle \tag{6.22}$$
$$U|10\rangle = |11\rangle \tag{6.23}$$
$$U|11\rangle = |10\rangle, \tag{6.24}$$

we have

$$U = \begin{pmatrix} 1 & 0 & 0 & 0 \\ 0 & 1 & 0 & 0 \\ 0 & 0 & 0 & 1 \\ 0 & 0 & 1 & 0 \end{pmatrix}. \tag{6.25}$$

The $n$-qubit state is represented by a vector of $2^n$ elements.

## 6.2 Controlled-operation gates

For two-qubit gates, the controlled-operation gate with a single-bit unitary transformation is defined as,

- If $A = 0, \rightarrow A' = A = 0$ and $B' = B$;
- If $A = 1, \rightarrow A' = A = 1$ and $B$ is transformed by $U$ as $B' = UB$.

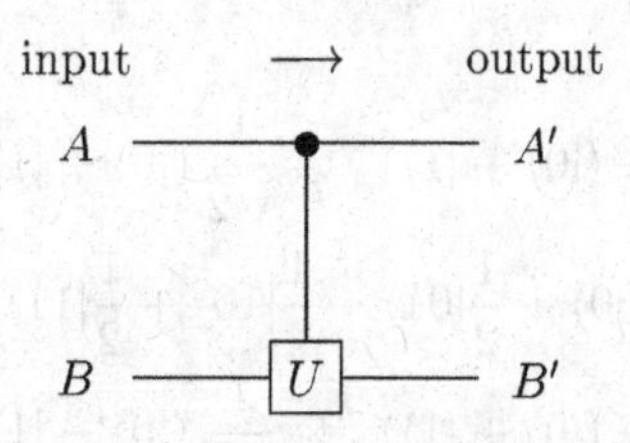


Fig. 6.6 Controlled-Operation Gate.

The circuit is illustrated in Fig. 6.6. The matrix becomes

$$\begin{pmatrix} 1 & 0 & 0 & 0 \\ 0 & 1 & 0 & 0 \\ 0 & 0 & \langle 0|U|0\rangle & \langle 0|U|1\rangle \\ 0 & 0 & \langle 1|U|0\rangle & \langle 1|U|1\rangle \end{pmatrix}. \tag{6.26}$$

In particular, in the case of the phase gate, $U = U(\phi)$ (see Section 6.1), the circuit is called the controlled-phase. This is one of the most important gates and applied to the quantum Fourier transformation.

$$U(\phi) = \begin{pmatrix} 1 & 0 & 0 & 0 \\ 0 & 1 & 0 & 0 \\ 0 & 0 & 1 & 0 \\ 0 & 0 & 0 & e^{i\phi} \end{pmatrix} \tag{6.27}$$

---

**Example 6.3** Express two circuits in Fig. 6.7 in terms of matrices.

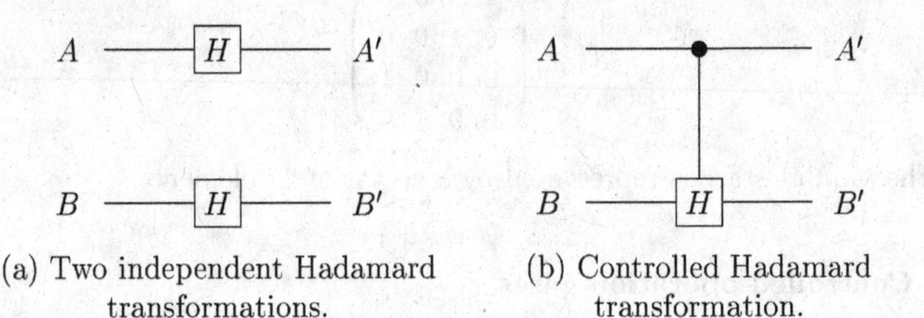


(a) Two independent Hadamard transformations.

(b) Controlled Hadamard transformation.

Fig. 6.7 Hadamard gates.

**Solution**

(a) Two Hadamard transformation: Let us denote the Hadamard operation as $O$. The states $|00\rangle$, $|01\rangle$, $|10\rangle$ and $|11\rangle$ are transformed by Hadamard transformations as

$$\begin{aligned}
O|00\rangle &= \frac{1}{\sqrt{2}}(|0\rangle + |1\rangle)_A \otimes \frac{1}{\sqrt{2}}(|0\rangle + |1\rangle)_B \\
&= \frac{1}{2}|00\rangle + \frac{1}{2}|01\rangle + \frac{1}{2}|10\rangle + \frac{1}{2}|11\rangle, \\
O|01\rangle &= \frac{1}{\sqrt{2}}(|0\rangle + |1\rangle)_A \otimes \frac{1}{\sqrt{2}}(|0\rangle - |1\rangle)_B \\
&= \frac{1}{2}|00\rangle - \frac{1}{2}|01\rangle + \frac{1}{2}|10\rangle - \frac{1}{2}|11\rangle, \\
O|10\rangle &= \frac{1}{\sqrt{2}}(|0\rangle - |1\rangle)_A \otimes \frac{1}{\sqrt{2}}(|0\rangle + |1\rangle)_B \\
&= \frac{1}{2}|00\rangle + \frac{1}{2}|01\rangle - \frac{1}{2}|10\rangle - \frac{1}{2}|11\rangle, \\
O|11\rangle &= \frac{1}{\sqrt{2}}(|0\rangle - |1\rangle)_A \otimes \frac{1}{\sqrt{2}}(|0\rangle - |1\rangle)_B \\
&= \frac{1}{2}|00\rangle - \frac{1}{2}|01\rangle - \frac{1}{2}|10\rangle + \frac{1}{2}|11\rangle.
\end{aligned}$$

Thus matrix elements of $O$ will be

$$O = \frac{1}{2}\begin{pmatrix} 1 & 1 & 1 & 1 \\ 1 & -1 & 1 & -1 \\ 1 & 1 & -1 & -1 \\ 1 & -1 & -1 & 1 \end{pmatrix}.$$

(b) Controlled Hadamard transformation: The transformation is given as

$$\begin{aligned} O|00\rangle &= |00\rangle, \\ O|01\rangle &= |01\rangle, \\ O|10\rangle &= |1\rangle_A \otimes \frac{1}{\sqrt{2}}(|0\rangle + |1\rangle)_B \\ &= \frac{1}{\sqrt{2}}|10\rangle + \frac{1}{\sqrt{2}}|11\rangle, \\ O|11\rangle &= |1\rangle_A \otimes \frac{1}{\sqrt{2}}(|0\rangle - |1\rangle)_B \\ &= \frac{1}{\sqrt{2}}|10\rangle - \frac{1}{\sqrt{2}}|11\rangle. \end{aligned}$$

Thus, we have

$$O = \begin{pmatrix} 1 & 0 & 0 & 0 \\ 0 & 1 & 0 & 0 \\ 0 & 0 & 1/\sqrt{2} & 1/\sqrt{2} \\ 0 & 0 & 1/\sqrt{2} & -1/\sqrt{2} \end{pmatrix}.$$

---

Similarly, the matrix representations of several gates are given as

- Swap
  exchanges two quantum bits. The matrix is

$$\begin{pmatrix} 1 & 0 & 0 & 0 \\ 0 & 0 & 1 & 0 \\ 0 & 1 & 0 & 0 \\ 0 & 0 & 0 & 1 \end{pmatrix}, \tag{6.28}$$

  and the circuit symbol is shown in Fig. 6.8. In classical computer, the exchange gate means nothing but the change of circuits. In quantum computers, however, each bit chronologically follows the evolution of state so that the exchange becomes a significant gate.

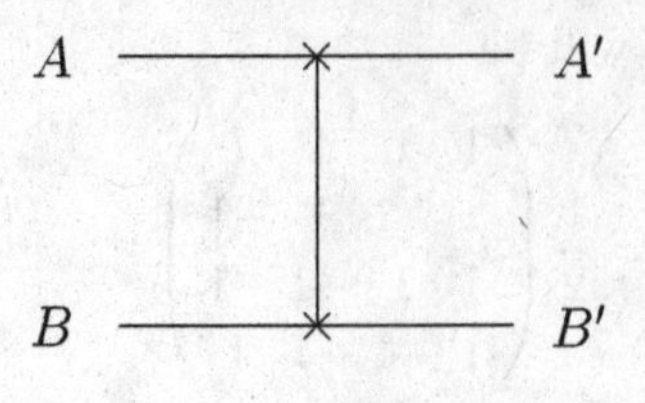


Fig. 6.8 Swap gate.

- Fredkin gate
  is a three-qubit gate called the controlled-exchange gate. Its operation is expressed as

$$A' = A \tag{6.29}$$
$$B' = (\overline{A} \cdot B) + (A \cdot C) \tag{6.30}$$
$$C' = (\overline{A} \cdot C) + (A \cdot B). \tag{6.31}$$

where $A$ is the controlled gate and $B'$ and $C'$ follow rules,

- If $A = 0$, the outputs $B'$ and $C'$ are not changed.
- If $A = 1$, the inputs $B$ and $C$ are exchanged to be the outputs of $C'$ and $B'$.

The truth table and the circuit symbol are shown in Fig. 6.9. Let us look at one output $B'$. If $A = 0$, the output is $B' = B$, while if $A = 1$ the output becomes $B' = C$. Thus it operates as a multiplexer with the address specified by $A$. The Fredkin gate can be used as other gates under specific conditions on the inputs. For example,

| $A$ | $B$ | $C$ | $A'$ | $B'$ | $C'$ |
|---|---|---|---|---|---|
| 0 | 0 | 0 | 0 | 0 | 0 |
| 0 | 0 | 1 | 0 | 0 | 1 |
| 0 | 1 | 0 | 0 | 1 | 0 |
| 0 | 1 | 1 | 0 | 1 | 1 |
| 1 | 0 | 0 | 1 | 0 | 0 |
| 1 | 0 | 1 | 1 | 1 | 0 |
| 1 | 1 | 0 | 1 | 0 | 1 |
| 1 | 1 | 1 | 1 | 1 | 1 |

The truth table

A A′
B B′
C C′

The circuit symbol

Fig. 6.9 Fredkin gate.

- AND gate by Fredkin gate: With the fixed value of $C = 0$, it acts as the gate AND, i.e.,

$$C' = A \cdot B.$$

- NOT gate by Fredkin gate: With the fixed values of $B = 0$ and $C = 1$, NOT gate is implemented as

$$C' = \overline{A}$$

A three-qubit state is represented by a column vector of $2^3 = 8$ elements and the operator is represented by $8 \times 8$ matrix. If we take the basis states of $|000\rangle$, $|001\rangle$, $|010\rangle$, $|011\rangle$, $|100\rangle$, $|101\rangle$, $|110\rangle$, $|111\rangle$ in this order, the matrix corresponding to the Fredkin gate is expressed as

$$\begin{pmatrix} 1 & 0 & 0 & 0 & 0 & 0 & 0 & 0 \\ 0 & 1 & 0 & 0 & 0 & 0 & 0 & 0 \\ 0 & 0 & 1 & 0 & 0 & 0 & 0 & 0 \\ 0 & 0 & 0 & 1 & 0 & 0 & 0 & 0 \\ 0 & 0 & 0 & 0 & 1 & 0 & 0 & 0 \\ 0 & 0 & 0 & 0 & 0 & 0 & 1 & 0 \\ 0 & 0 & 0 & 0 & 0 & 1 & 0 & 0 \\ 0 & 0 & 0 & 0 & 0 & 0 & 0 & 1 \end{pmatrix} \tag{6.32}$$

---

**Example 6.4** Show that the OR gate is constructed by a combination of AND and NOT gates.

**Solution** Figure 6.10 shows an example.

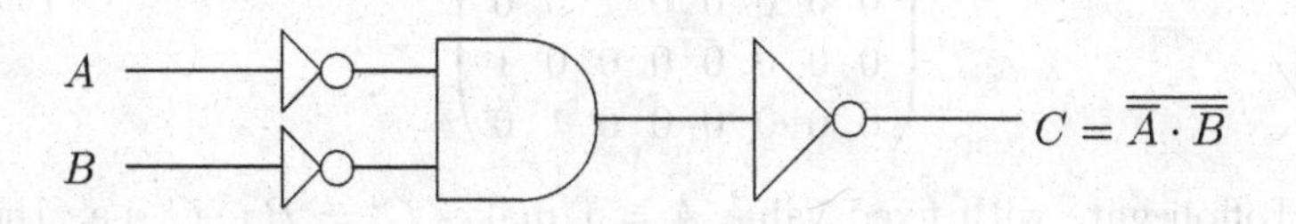


Fig. 6.10 OR circuit by a combination of AND and NOT gates.

---

In this way, the Fredkin gate can act as AND, OR and NOT gates so that any logical circuit can be constructed by a combination of the Fredkin gates.

- Toffoli gate

| $A$ | $B$ | $C$ | $A'$ | $B'$ | $C'$ |
|---|---|---|---|---|---|
| 0 | 0 | 0 | 0 | 0 | 0 |
| 0 | 0 | 1 | 0 | 0 | 1 |
| 0 | 1 | 0 | 0 | 1 | 0 |
| 0 | 1 | 1 | 0 | 1 | 1 |
| 1 | 0 | 0 | 1 | 0 | 0 |
| 1 | 0 | 1 | 1 | 0 | 1 |
| 1 | 1 | 0 | 1 | 1 | 1 |
| 1 | 1 | 1 | 1 | 1 | 0 |

The truth table

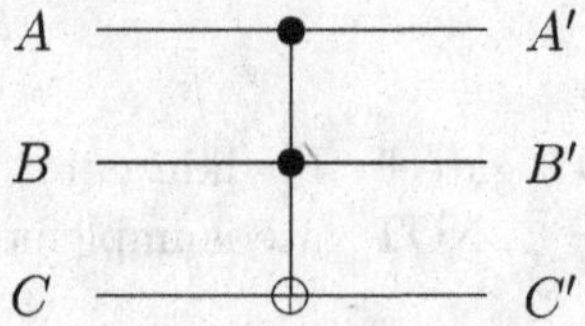

The circuit symbol

Fig. 6.11 Toffoli gate.

is the Controlled-Controlled-NOT gate defined by

$$A' = A \tag{6.33}$$

$$B' = B \tag{6.34}$$

$$C' = C \oplus (A \cdot B) \tag{6.35}$$

The truth table and the circuit symbol are shown in Fig. 6.11. The operation is represented by a matrix:

$$\begin{pmatrix} 1 & 0 & 0 & 0 & 0 & 0 & 0 & 0 \\ 0 & 1 & 0 & 0 & 0 & 0 & 0 & 0 \\ 0 & 0 & 1 & 0 & 0 & 0 & 0 & 0 \\ 0 & 0 & 0 & 1 & 0 & 0 & 0 & 0 \\ 0 & 0 & 0 & 0 & 1 & 0 & 0 & 0 \\ 0 & 0 & 0 & 0 & 0 & 1 & 0 & 0 \\ 0 & 0 & 0 & 0 & 0 & 0 & 0 & 1 \\ 0 & 0 & 0 & 0 & 0 & 0 & 1 & 0 \end{pmatrix} \tag{6.36}$$

The Toffoli gate with fixed value $A = 1$ makes $C' = B \oplus C$, i.e., the controlled-NOT. The fixed value $C = 0$ makes AND, i.e., $C' = A \cdot B$.

---

**Example 6.5** Design a half adder and a full adder by a combination of the Toffoli gates and the controlled-NOT gates.

**Solution**

Half adder: A logical sum of inputs $X$ and $Y$ is expressed as $S = X \oplus Y$ and can be attained by a controlled-NOT gate. The carry $C = X \cdot Y$ is also implemented by a controlled-controlled-NOT gate with the fixed third input fixed to be 0. Thus the circuit in Fig. 6.12(a) works as a half adder.

Full adder: The sum of inputs $X \oplus Y \oplus Z$ can be made by two controlled-controlled-NOT gates. Including the carry part, $C = X \cdot Y + Y \cdot Z + Z \cdot X$, the circuit is shown in Fig. 6.12(b).

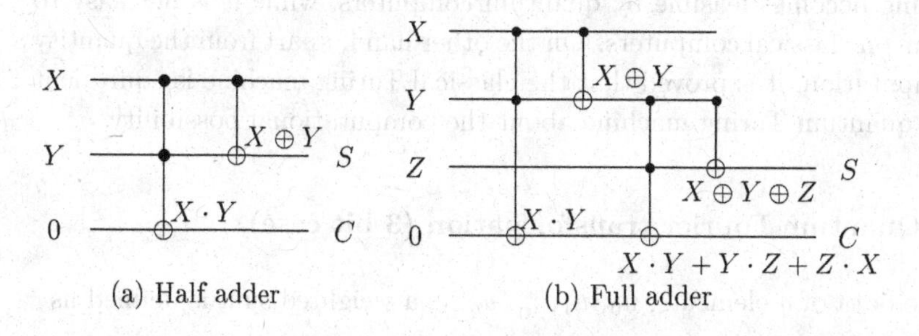


Fig. 6.12 Adders by reversible logic gates.

## 6.3 Quantum Turing machine

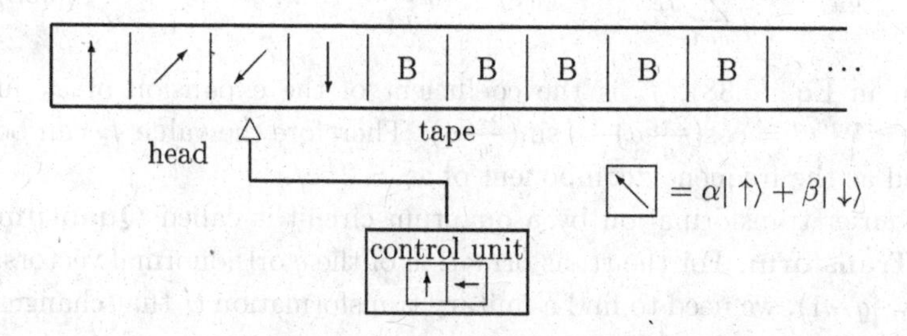


Fig. 6.13 Quantum Turing machine.

Quantum Turing machine was proposed by C. H. Bennett, P. Benioff and R. P. Feynman as a quantum-mechanical version of the Turing machine. One of the well-defined machines is given by D. Deutsch (Fig. 6.13). Each cell can take a superposition of 0 and 1. The upward arrow stands for 0 while the downward one means 1. The declined arrow corresponds to a qubit with superposition of 0 and 1. There is a notation which introduces a three-dimensional sphere to express a phase of superposed states.

The quantum Turing machine looks like a kind of classical non-deterministic or probabilistic Turing machine. The classical machines can choose only single process, but the quantum Turing machine can perform parallel computation based on the quantum mechanical superposition. With this unique character, a certain kind of computations including data searching becomes feasible by quantum computers, while it is not easy to perform on classical computers. On the other hand, apart from the quantity of computation, it is proved that the classical Turing machine is equivalent to the quantum Turing machine about the computational possibility.

## 6.4 Quantum Fourier transformation (3 bit case)

For the data of $q$ elements, $u_0$, $u_1$, ..., $u_{q-1}$, a weighted sum is defined as

$$f_c = \frac{1}{\sqrt{q}} \sum_{a=0}^{q-1} u_a \, e^{2\pi i a c/q} \qquad (c = 0,\, 1,\, ...,\, q-1). \tag{6.37}$$

where the set of $q$ numbers, $f_0$, $f_1$, ..., $f_{q-1}$, is called the **discrete Fourier transform** of $u_0$, $u_1$, ..., $u_{q-1}$. On the contrary, if the Fourier transform is given, the original data can be recovered by the inverse transformation:

$$u_a = \frac{1}{\sqrt{q}} \sum_{c=0}^{q-1} f_c \, e^{-2\pi i a c/q} \qquad (a = 0,\, 1,\, ...,\, q-1). \tag{6.38}$$

As is seen in Eq. (6.38), $f_c$ is the coefficient of the expansion of $u_a$ in terms of $e^{-2\pi i a c/q} = \cos(\frac{2\pi c}{q} a) - i \sin(\frac{2\pi c}{q} a)$. Therefore the value $f_c$ can be interpreted as the frequency component of $\omega_c = 2\pi c/q$.

The Fourier transformation by a quantum circuit is called **Quantum Fourier Transform**. For the transformation of the $q$ orthonormal vectors, $|0\rangle$, $|1\rangle$, ..., $|q-1\rangle$, we need to find a unitary transformation $U$ that changes the state $|a\rangle$ $(0 \leq a \leq q-1)$ into

$$\frac{1}{\sqrt{q}} \sum_{c=0}^{q-1} e^{2\pi i a c/q} |c\rangle. \tag{6.39}$$

With this transformation, a state vector of the components: $u_0$, $u_1$, $u_2$, ..., $u_{q-1}$ in the basis $|a\rangle$ will be transformed into a vector with the components $f_0$, $f_1$, $f_2$, ..., $f_{q-1}$. In this way, the original state vector

$$|A\rangle = \sum_a u_a |a\rangle, \tag{6.40}$$

is transformed into $|a\rangle$ by (6.39)

$$|B\rangle = U|A\rangle = \sum_a u_a \left( \frac{1}{\sqrt{q}} \sum_c e^{2\pi iac/q} |c\rangle \right)$$
$$= \sum_c \left( \frac{1}{\sqrt{q}} \sum_a u_a e^{2\pi iac/q} |c\rangle \right), \qquad (6.41)$$

by Eq. (6.39) so that the components of $|c\rangle$ are equal to $f_c$ defined by Eq. (6.37).

As an example of the transformation (6.39), let us perform the Fourier transformation with $q = 2^3 = 8$. With three bits: $|q_2q_1q_0\rangle$, Each of the three bits $a_2a_1a_0$ for the binary representation of an integer $a = 2^0a_0+2^1a_1+2^2a_2$ is assigned to the corresponding qubit of the state $|q_2q_1q_0\rangle$. We set also the binary representation for the components $c$ after the transformation; $c = 2^0c_0 + 2^1c_1 + 2^2c_2$, Noting $e^{2\pi i \times \text{integer}} = 1$, we have

$$|a_2a_1a_0\rangle \to \frac{1}{\sqrt{8}} \sum_{c_2=0}^{1}\sum_{c_1=0}^{1}\sum_{c_0=0}^{1} e^{2\pi i(a_0+2a_1+4a_2)(c_0+2c_1+4c_2)/8} |c_2c_1c_0\rangle$$
$$= \left( \frac{1}{\sqrt{2}} \sum_{c_2=0}^{1} e^{2\pi i(4c_2)(a_0+2a_1+4a_2)/8} |c_2\rangle_2 \right)$$
$$\otimes \left( \frac{1}{\sqrt{2}} \sum_{c_1=0}^{1} e^{2\pi i(2c_1)(a_0+2a_1+4a_2)/8} |c_1\rangle_1 \right)$$
$$\otimes \left( \frac{1}{\sqrt{2}} \sum_{c_0=0}^{1} e^{2\pi i c_0(a_0+2a_1+4a_2)/8} |c_0\rangle_0 \right)$$
$$= \frac{1}{\sqrt{2}} \left( |0\rangle_2 + e^{2\pi i \cdot a_0/2} |1\rangle_2 \right)$$
$$\otimes \frac{1}{\sqrt{2}} \left( |0\rangle_1 + e^{2\pi i \cdot (a_1/2+a_0/4)} |1\rangle_1 \right)$$
$$\otimes \frac{1}{\sqrt{2}} \left( |0\rangle_0 + e^{2\pi i \cdot (a_2/2+a_1/4+a_0/8)} |1\rangle_0 \right). \qquad (6.42)$$

Now let us try to make a circuit to implement the transformation (6.42). For the first line of the right-hand side, one should note that

- $\frac{1}{\sqrt{2}}(|0\rangle + |1\rangle)$ if $a_0 = 0$,
- $\frac{1}{\sqrt{2}}(|0\rangle - |1\rangle)$ if $a_0 = 1$,

which is equivalent to the Hadamard transformation on $|a_0\rangle$ as shown in Fig. 6.14. This circuit generates $|c_2\rangle$. In the second line, the $a_1$ dependence

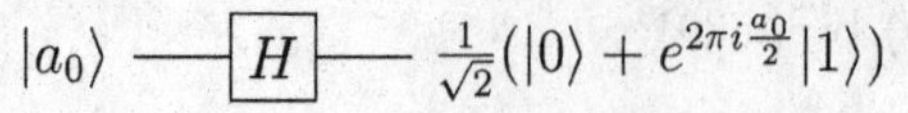

$$|a_0\rangle \;—\boxed{H}—\; \frac{1}{\sqrt{2}}(|0\rangle + e^{2\pi i \frac{a_0}{2}}|1\rangle)$$

Fig. 6.14 Circuit for $a_0$.

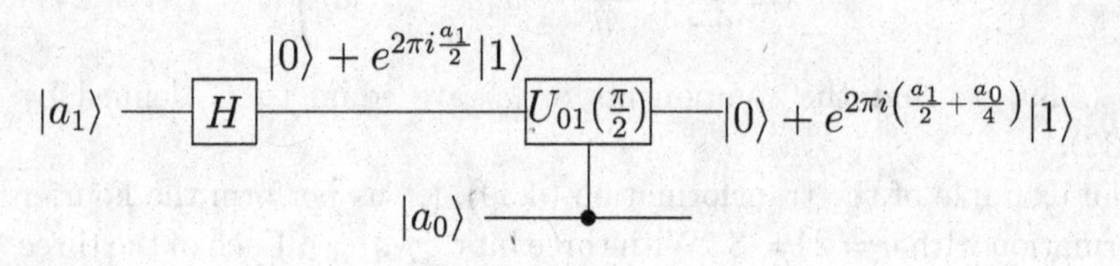


Fig. 6.15 Circuit from $a_1$.

is implemented also by the Hadamard gate on $|a_1\rangle$. For $a_0$ dependence of the second line, note that

- $e^{2\pi i \cdot a_0/4} = 1$ if $a_0 = 0$,
- $e^{2\pi i \cdot a_0/4} = e^{i\pi/2}$ $(= i)$ if $a_0 = 1$.

Thus this can be implemented by applying the controlled-phase gate $U(\phi = \pi/2)$ of Section 6.1. This circuit generates the bit $c_1$. Finally, the last line is implemented by Fig. 6.16 for the output $c_0$. To summarize three circuits for the outputs $c_0$, $c_1$ and $c_2$, we obtain the Fourier transformation circuit for three qubits ($q = 8$) shown in Fig. 6.17.

Generalization ($q = 2^n$ A$n \geq 4$) is straightforward and will be discussed in Chapter 6.

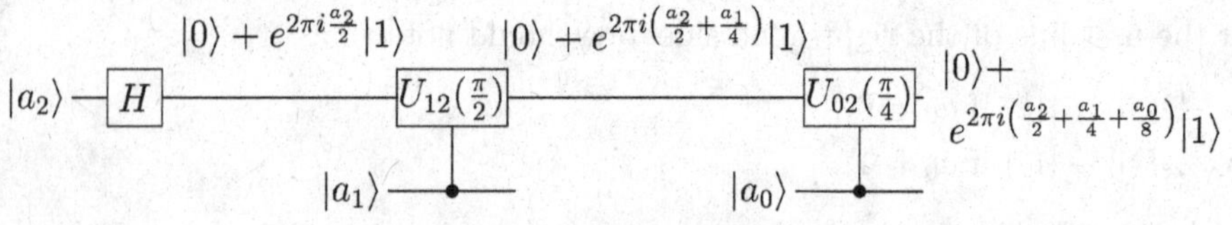

$|a_2\rangle$ — $H$ — $|0\rangle + e^{2\pi i \frac{a_2}{2}}|1\rangle$ — $U_{12}(\frac{\pi}{2})$ (controlled by $|a_1\rangle$) — $|0\rangle + e^{2\pi i(\frac{a_2}{2}+\frac{a_1}{4})}|1\rangle$ — $U_{02}(\frac{\pi}{4})$ (controlled by $|a_0\rangle$) — $|0\rangle + e^{2\pi i(\frac{a_2}{2}+\frac{a_1}{4}+\frac{a_0}{8})}|1\rangle$

Fig. 6.16 Circuit for $a_2$.

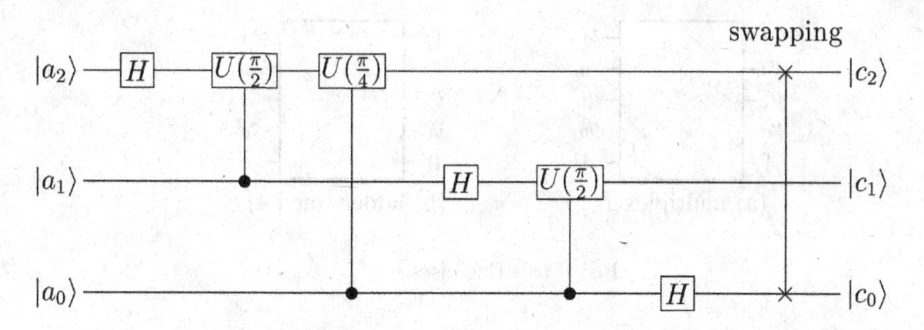


Fig. 6.17 Circuit for three qubit Fourier transformation.

**Summary of Chapter 6**

(1) Basic operations in quantum computers are implemented by quantum gates, which perform unitary transformations on qubits. In particular, an Hadamard gate is most commonly used gate for quantum circuits.
(2) Controlled-operation gates play important roles in quantum circuits to implement Unitary transformations and Boolean algebras.
(3) Quantum Turing machine is a quantum mechanical version of Turing machine which performs parallel computations with qubits.
(4) The Fourier transformation of qubits can be performed by a fast quantum algorithm known as the Quantum Fourier transformation.

## 6.5 Problems

[1] For the Fredkin gate and the Toffoli gate, derive logical equations which express the inputs $(A, B, C)$ in terms of the outputs $(A', B', C')$.

[2] Design a two bit multiplexer combining Fredkin gates. Specifically, design a circuit; for the inputs $X = x_1x_0$, $Y = y_1y_0$ and the address bit $A$, the output is $Z = z_1z_0 = X$ if $A = 0$ and $Z = Y$ if $A = 1$, as shown in Fig. 6.18(a).

[3] Design an OR circuit using the Fredkin gates only.

[4] Design an adder of mod 4 by the controlled-NOT and the Toffoli gates which performs $Z = X + Y \pmod 4$ for two bit numbers $X = x_1x_0$, $Y = y_1y_0$ (see Fig. 6.18(b)).

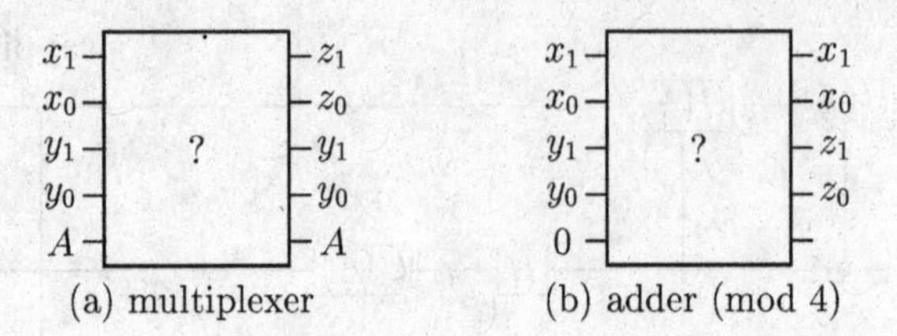


Fig. 6.18 Exercises.

[**5**] Show that the discrete Fourier transformation (6.37) is a unitary transformation. Show also that the original values $u_a$ are obtained by summing $f_c$ in Eq. (6.37).

[**6**] Verify that the circuit of Fig. 6.17 performs the transformation (6.39), by multiplying $8 \times 8$ matrices corresponding to the gates in the figure.

# Chapter 7

# Theory of Information and Communication

The concept of **entropy** was firstly introduced in thermodynamics to measure the disorder and the uncertainty of a system. In physics, the entropy has been used to measure the diffusion of heat or materials. Nowadays, the idea of entropy is used also in the communication theory as a measure of the transmission speed of information, i.e., how much information can be delivered in a fixed time period. In data communication, the information is transmitted as bits, irrespective of whether it contain characters or pictures. There are many algorithm to change the data to the bit information. If the amount of contents is the same, it is obvious that the less bit length makes the faster communication. Thus, the concept of entropy is closely related to the problem of data compression.

In this chapter, we describe firstly the classical theory of information where the Shannon's theory is important. Then as the extension of the classical entropy to the quantum information, we introduce the von Neumann's entropy.

**Keywords: entropy (Shannon Entropy); source coding theorem (Shannon's first theorem); von Neumann entropy; channel coding theorem (Shannon's second theorem).**

## 7.1 Shannon entropy

### 7.1.1 *Amount of information*

In a given system, if all $N$ events occur with equal probabilities, the amount of information obtained by these events is defined as

$$I = \log_2 N \tag{7.1}$$

in units of bit. For example, if you throw a coin, the amount of information is $I = \log_2 2 = 1$ because $N = 2$ for the head and tail of a coin. If the coin is thrown twice, the information of knowing the result of both trials is $I = \log_2 2^2 = 2$ bits because there are $2^2 = 4$ possibilities: "head-head", "head-tail", "tail-head" and "tail-tail". Similarly, if the coin is thrown three times, the information is 3 bits. Thus, Eq. (7.1) looks a reasonable definition to understand intuitively the amount of information of events.

In general, if the first and second events occur independently with $N_1$ and $N_2$ possibilities, the observation of two events will determine one possible event among $N_1 \times N_2$ possibilities. Therefore the amount of information is $I_{12} = \log_2(N_1 \times N_2)$. This exactly agrees with the sum of the amount of information of the first event $I_1 = \log_2 N_1$ and the second $I_2 = \log_2 N_2$.

---

**Example 7.1** How much is the amount of information of throwing a dice?

**Solution** Since a dice has six numbers, the amount of information is $\log_2 6 \approx 2.525$ bits.

---

### 7.1.2 *Entropy*

When characters are sent successively, what will be the amount of information per one character? The difference from the cases with coins and dice is that the probabilities of having what characters are not equal. Let us assume that there are $N$ kinds of characters and the probability of concurrence of $i$th character is $p_i$. For example, the alphabet has $N = 26$ characters apart from some special characters like period, colon and semicolon. Though we are thinking "characters" now, other cases like the head or tail of a coin, the number of a dice can be treated in a similar way. As the amount of information per a single character, Claude E. Shannon defined it a formula

$$I = -\sum_{i=1}^{N} p_i \log_2 p_i, \tag{7.2}$$

and called it **entropy** (the **Shannon entropy**). Equation (7.2) can be interpreted as the result of averaging the entropy of a single character

$$I_i = -\log_2 p_i \tag{7.3}$$

over all characters. For $p_i = 0$, the logarithm is not defined so that the limit $\lim_{p_i \to 0} p_i \log_2 p_i = 0$ is adopted. Since $0 \leq p_i \leq 1$, each term of Eq. (7.2) is positive: $-p_i \log_2 p_i \geq 0$ and $I \geq 0$. The entropy (7.2) is a generalization of the amount of information (7.1). In fact, if the $N$ characters appear with the equal probability $p_i = 1/N$ for all $i$, the entropy becomes

$$I = -\sum_{i=1}^{N} \frac{1}{N} \log_2 \frac{1}{N} = \log_2 N$$

and agrees with (7.1).

In the case of two different results $i = 1$ and $i = 2$, let us assume that $N_1$ events for $i = 1$, and $N_2$ events for $i = 2$ among the total events $N = N_1 + N_2$. The amount of information is obtained by (7.1) with all possible combinations of events $N!/N_1!N_2!$ as

$$I = \log_2 \frac{(N_1 + N_2)!}{N_1!N_2!}. \tag{7.4}$$

Using the Stirling's formula at large $N$ limit,

$$\ln(N!) \approx N \ln N - N \tag{7.5}$$

and $\log_2 N! = \log_2 e \times \ln N!$, we can rewrite Eq. (7.4) approximately to be

$$I \approx -(N_1 + N_2) \left[ \frac{N_1}{N_1 + N_2} \log_2 \frac{N_1}{N_1 + N_2} + \frac{N_2}{N_1 + N_2} \log_2 \frac{N_2}{N_1 + N_2} \right]. \tag{7.6}$$

This equation (7.6) agrees with the entropy per character (7.2) because $p_i = N_i/(N_1 + N_2)$.

---

**Example 7.2** If the probabilities of events 1, 2 and 3 are $p_1$, $p_2$ and $p_3$, respectively, show that the entropy $I$ of the system is given by

$$I(p_1, p_2, p_3) = I(p_1, p_2 + p_3) + (p_2 + p_3) I \left( \frac{p_2}{p_2 + p_3}, \frac{p_3}{p_2 + p_3} \right) \tag{7.7}$$

where $I(p_1, p_j) = p_i \log p_i + p_j \log p_j$ .

**Solution** From the definition (7.2), we have

$$
\begin{aligned}
&I(p_1, p_2, p_3) \\
&= -\left[p_1 \log p_1 + p_2 \log p_2 + p_3 \log p_3\right] \\
&= -p_1 \log p_1 \\
&\quad - \left[+p_2 \log\left((p_2+p_3)\cdot\frac{p_2}{p_2+p_3}\right) + p_3 \log\left((p_2+p_3)\cdot\frac{p_3}{p_2+p_3}\right)\right] \\
&= -p_1 \log p_1 - \left[+p_2\left(\log(p_2+p_3) + \log\frac{p_2}{p_2+p_3}\right)\right. \\
&\quad \left. + p_3\left(\log(p_2+p_3) + \log\frac{p_3}{p_2+p_3}\right)\right] \\
&= -\left[p_1 \log p_1 + (p_2+p_3)\log(p_2+p_3)\right] \\
&\quad - \left[p_2 \log\frac{p_2}{p_2+p_3} + p_3 \log\frac{p_3}{p_2+p_3}\right] \\
&= -\left[p_1 \log p_1 + (p_2+p_3)\log(p_2+p_3)\right] \\
&\quad - (p_2+p_3)\left[\frac{p_2}{p_2+p_3}\log\frac{p_2}{p_2+p_3} + \frac{p_3}{p_2+p_3}\log\frac{p_3}{p_2+p_3}\right] \\
&= I(p_1, p_2+p_3) + (p_2+p_3) I\left(\frac{p_2}{p_2+p_3}, \frac{p_3}{p_2+p_3}\right).
\end{aligned}
\tag{7.8}
$$

This result means that the entropy $I(p_1, p_2, p_3)$ of the events 1, 2 and 3 agrees with the sum of two terms: the first term $I(p_1, p_2 + p_3)$ corresponds to the entropy with the alternative of the event 1 or "one event from 2 or 3"; the second $I(p_2/(p_2+p_3), p_3/(p_2+p_3))$ corresponds to the entropy with the alternative of 2 or 3 under the condition that "2 or 3" is already chosen with the weight $(p_2 + p_3)$.

---

### 7.1.3 *Coding information*

The transformation of characters, pictures and symbols into a set of simple symbols is called **coding**. The set of symbols obtained is called **codes**. Usually it looks like a kind of pattern so that it is called the bit pattern. Let us now consider how one can make the code efficiently.

---

**Example 7.3** Let us consider the case that 200 students attend a lecture and a teacher evaluates their achievement. The marks are recorded by ranks: A (excellent), B (good), C (average) and F (failure), with probabilities 1/2, 1/4, 1/8 and 1/8, respectively. Make the code of records into the combinations of 0 and 1 with the total length as short as possible. What is the proper way of assigning the bit patterns made of0 and 1 to A, B, C and F?

**Solution** If you use two bits, there are four combinations which can be assigned to the ranks as

$$A \to 00$$

$$B \to 01$$

$$C \to 10$$

$$F \to 11.$$

For example,the bit pattern

$$0001111000000110...$$

can be read as

$$ABFCAABC....$$

Now the question is that the above solution is the best one? The total length $200 \times 2 = 400$ bits is needed to store the data of the 200 students in the above solution. Is there any way to store them in less space? The answer is as follows. If one assign the following patters for A, B, C and F,

$$A \to 0$$

$$B \to 10$$

$$C \to 110$$

$$F \to 111,$$

the same data, ABFCAABF, is coded as

$$\overbrace{0}^{A}\ \overbrace{10}^{B}\ \overbrace{111}^{F}\ \overbrace{110}^{C}\ \overbrace{0}^{A}\ \overbrace{0}^{A}\ \overbrace{10}^{B}\ \overbrace{110}^{C}.$$

One can see easily that the original pattern ABF ... can be recovered from the above bit pattern.

By using the latter algorithm, the record for

- A ... 100 students
- B ... 50 students
- C ... 25 students
- F ... 25 students

can be stored in the bit length,

$$100 \times 1 + 50 \times 2 + 25 \times 3 + 25 \times 3 = 350 \text{ (bits)}.$$

so that the average bits per student has been reduced from 2 bits to 1.75 bits.

---

It is seen now that the former solution of Problem in Example 7.3 is not the best way to code the ranks of 200 students.

We may raise a naive question how much the total length of the bit pattern can be shortened for a given data. **The source coding theorem (Shannon's first theorem)** answers to this question. According to this theorem, the average size per symbol can be shortened up to the lower bound specified by the entropy. In other words, the average size cannot be shortened less than the entropy. Let us go back the above example of school records? The entropy is evaluated to be

$$H = \frac{1}{2}\log_2 2+\frac{1}{4}\log_2 4+\frac{1}{8}\log_2 8+\frac{1}{8}\log_2 8 = 0.5+0.5+0.375+0.375 = 1.75.$$

This entropy tells that the average bit size per student can be shortened up to 1.75 bits in the optimized way, The second coding scheme above thus turns out to be the optimal one.

You may doubt the above theorem if you consider the following example about the weather. Let us assume that the probability of rain is 80%, while that of no rain is 20%. The entropy is given for this weather probabilities as

$$0.8 \times \log_2 \frac{1}{0.8} + 0.2 \times \log_2 \frac{1}{0.2} = 0.722.$$

This entropy implies that the shortest bit size should be 0.722 bits for this weather information. If you assign "0" to "no rain" and "1" to "rain", the average bit length is just one! How can one cut short the average length? The answer to this question is to combine two or more data. For example, combine two data as "no rain - no rain", "no rain - rain", "rain - no rain" and "rain - rain". While two bits will be needed in the straightforward coding, let us give a short code to the pair with the highest probability and a long code for that with lower probability. By assigning the same codes as the previous example, we obtain the following table;

| | probability | symbol | length |
|---|---|---|---|
| ○ ○ → | 0.64 | 0 | 1 |
| ○ ● → | 0.16 | 10 | 2 |
| ● ○ → | 0.16 | 110 | 3 |
| ● ● → | 0.04 | 111 | 3 |
| | | | average 1.56 |

where "no rain" is represented by $\circ$ and "rain" is shown by $\bullet$. Since the average bits 1.56 is for two days, the average bits per day is 0.78 bits. This is still slightly larger than the lower limit of 0.722 bits, but there is an appreciable improvement compared to the average bit 1 by the simplest coding. As a compensation of short length, we sacrifice the simplicity of transformation from the original data to the codes. In general, it is rational to assign less bits to frequently used symbols and longer bits to rarely use symbols. This idea is developed further by the **Huffman code** in which the coding is systematically performed with this algorithm.

### 7.1.4 *Von Neumann entropy*

The Shannon entropy measures the uncertainty associated with a classical probability distribution. The entropy of quantum mechanical system will be formulated by using the **density operator** $\rho$. The term "**density matrix** is also used as the same meaning). The von Neumann entropy is defined by using the density operator as

$$S(\rho) = -\sum_n \langle n|\rho \log \rho|n\rangle \equiv -\mathrm{tr}(\rho \log \rho). \tag{7.9}$$

where the density matrix is given by

$$\rho = \sum_n p_n \, |n\rangle\langle n|, \tag{7.10}$$

for an ensemble of the quantum states $|n\rangle$ in a Hilbert space with respective probabilities $p_n$. The von Neumann entropy is then given by

$$S(\rho) = -\sum_n p_n \log p_n. \tag{7.11}$$

If the system is kept under a temperature $T$, the probability $p_n$ is proportional to the Boltzmann factor, $p_n \propto e^{-E_n/(k_{\mathrm{B}}T)}$ where $E_n$ is the eigenenergy of the state $|n\rangle$ and $k_{\mathrm{B}}$ is the Boltzmann constant.

## 7.2 Transmission error in communication

### 7.2.1 *Errors and channel capacity*

Let us think of sending some information. In their way to the destination, The signals may be distorted and disturbed by many kinds of origins in the way to the destination. It might be impossible to avoid receiving incorrect signals. The origins that cause errors irregularly are called **noise** of communication.

We try to send a set of information in the form of bits. If errors never happen, it is clear that one bit carries exactly one bit of information. The real situation is, however, not like that. For simplicity, we assume that the error occurs with probability $q$ per bit. It is the same as saying that the signal carrying "0" is received as "1" with probability $q$ and the signal carrying "1" is received as "0" with probability $q$. We also assume that the probability of error does not depend on the signals sent before.

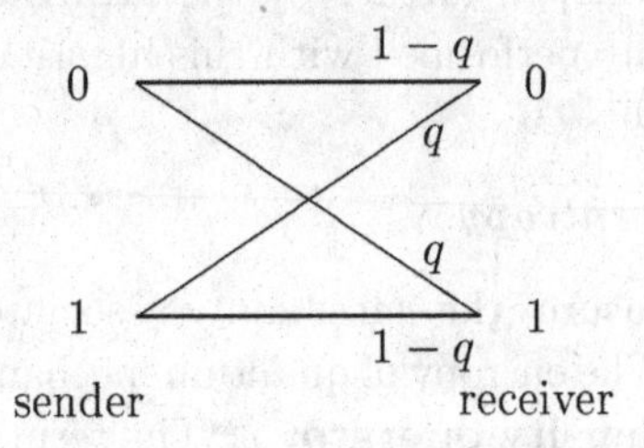


Fig. 7.1 Communication of binary signals under noise.

We will examine hereafter the case with $q = 1/5$; the wrong information is delivered at a rate of once in five times. Under this circumstance, let us try to send information with the error rate as low as possible. How about sending each signal three times?

- The sender sends each signal three times, namely, he sends 000 if the original value is 0, while he sends 111 for 1.
- If the three values do not agree, the receiver interprets the majority of the received signals to be the correct one.

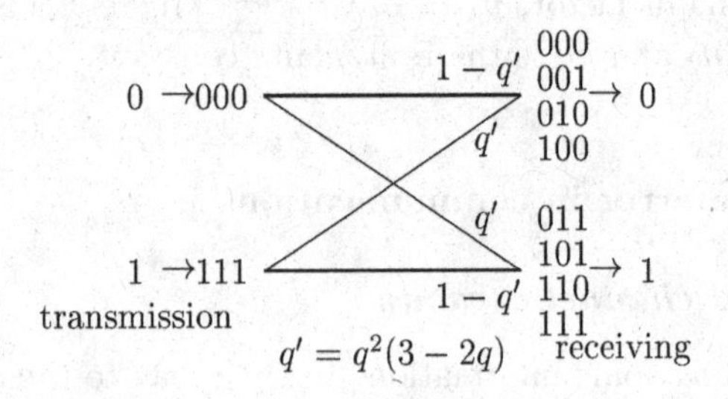


Fig. 7.2 Communication by binary signals under noise. If one sends the same signal three times, the probability of error is reduced.

In this method, the receiver gets error signals if more than two out of the three bits are inverted. This probability is given by

$$q' = q^3 + 3q^2(1-q) = q^2(3-2q). \tag{7.12}$$

In the case of $q = 1/5 = 0.2$, the probability of errors is reduced to $q' = 13/125 = 0.104$ by the method, while we have to pay the expense of "redundancy". In other words, it reduces the error at the cost of slow down the communication speed to one third. In this method, the error is corrected if it occurs only one of the three bits. But if two or three bits are errors, the receiver gets the incorrect information.

In comparing two bit patterns of the same length, **the Hamming distance** is defined as the number of bits where the two patterns have different values. For example, the Hamming distance between two patterns:

$$A = 1111000011110000$$
$$B = 0101000000011010$$

is seven, because the 1st, 3rd, 9th, 10th, 11th, 13th and 15th bits are different. The Hamming distance can be used to correct the error in the communication. Suppose we knew beforehand that one of the two patterns either $A$ or $B$ is the correct bits. When we receive the code by the transmission,

$$C = 0101000001110000,$$

one can judge the correct code among $A$ and $B$ comparing the Hamming distances. The Hamming distances are easily obtained to be three and four between the codes $C$ and $A$, and between $C$ and $B$, respectively. Then we can conclude that the bit pattern $A$ is the correct one since the Hamming distance is short.

In general, in order to reduce the error, messages are transformed into the codes with redundancy. **The channel coding theorem** (Shannon's second theorem) discusses how much redundancy of codes is necessary to minimize the error rate. According to the theorem, one can define the **channel capacity** inherent in any communication channel. Let us define the ratio $R =$ "the number of information bits"/"the number of bits of the code" for the communication channel. As long as the rate $R$ does not exceed the capacity $C$, $R \leq C$, one can make the coding with the minimal error rate. Contrarily, if $R > C$, there is a limit to decrease the error rate in all possible coding system.

In the problem mentioned in this section, the channel capacity is known to be

$$C = 1 - q\log_2\frac{1}{q} - (1-q)\log_2\frac{1}{1-q}. \tag{7.13}$$

For $q = 1/5$, $C \approx 0.278$. According to the theorem, there must be a coding system with infinitely small error probability if one makes the code system in which the amount of information per bit is smaller than 0.278 bits. If we send one signal three times with this channel, the ratio becomes $R = 1/3 \approx 0.333$, which is not small enough to guarantee the safe information transmission with the minimal error. In fact, the error rate $q'$=0.104 might not be enough for the secure transmission.

The speed of communication is defined as the amount of information transmitted per second, i.e., the speed is obtained by a product of the channel capacity $C$ and the number of bits sent per second. For a system which can send 100,000 bits per second and the communication capacity is $C \approx 0.278$, i.e., $q = 1/5$ in Eq. (7.13), the speed is 27,800 bits per second. Then, one can make a code with infinitesimal error rate if one restrict the transmitted speed less than 27,800 bits per second.

### 7.2.2 *Shannon's theorem*

Let us suppose a channel which error rate per bit is $q < 1/2$. We would like to send the data $M$ bits in that channel. However, $qM$ bits will be errors. The receiver cannot recognize which are error bits. To avoid such a problem, the sender adds redundant bits and makes the code having the total $L$-bit length. In this case, the number of error bits is $qL$ in average. The actual number of the error bits may distribute around the average number. Let us write the probability of having $n$ error bits as $P(n)$. The value $P(n)$ can be obtained by simple combinatorics as a binomial distribution,

$$P(n) = q^n(1-q)^{L-n}\binom{L}{n}, \tag{7.14}$$

where $\binom{L}{n} = L!/((L-n)!n!)$ is the binomial coefficients. Figure 7.3 (a) shows the probability $P(n)$ for the number of error bits $n$ out of total number $L = 10$ with $q = 0.3$. The probability reaches its maximum at its average $n = 3$. But it distributes also around this value. For example, there is appreciable probability of having six error bits. Now, let us look at the behavior of $P(n)$ as $L$ gets larger, keeping $q = 0.3$ fixed. Figure 7.3 (b) shows the case with $L = 1000$. The probability $P(n)$ has the maximum at $n = qL = 300$, and the number of errors distributes very tightly around the maximum $\pm 50$ as seen. Out of that this range, the probability is very small. This tendency becomes stronger with larger $L$, as easily verified.

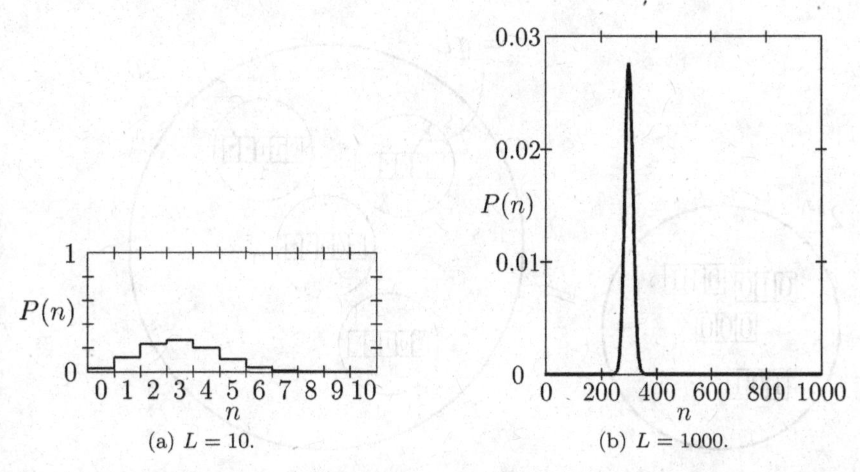


(a) $L = 10$. (b) $L = 1000$.

Fig. 7.3 Distribution of the number of error bits ($q = 0.3$). The two figures are scaled so that the areas below the graphs are equal (=1).

With sufficiently large $L$, the information can be delivered with very high accuracy if the code corrects the errors up to $qL$ bits out of the total $L$ bits. For that purpose, the coding system should be made so that there is only one correct code which can be reached by the correction of $qL$ bits. The number of codes which can be reached within the distance of $qL$ is

$$V \approx \sum_{n=0}^{qL} \binom{L}{n}. \tag{7.15}$$

If $L$ is large enough, the number $\binom{L}{n}$ increases as a function of $n$ quite rapidly when $n < L/2$. Therefore, the sum is almost determined by the last point $n = qL$: $\binom{L}{qL}$, therefore we have

$$V \approx \binom{L}{qL}. \tag{7.16}$$

We should arrange the codes so that different correct combinations do not have any overlap. Figure 7.4 schematically shows the situation. Paying attention to "The number of different messages"$=2^M$, we note that this number must be smaller than the number of "the total number of combinations"$=2^L$ divided by "the number of combinations that can be reached by mistake"$= \binom{L}{qL}$. Otherwise, the receiver cannot tell which is

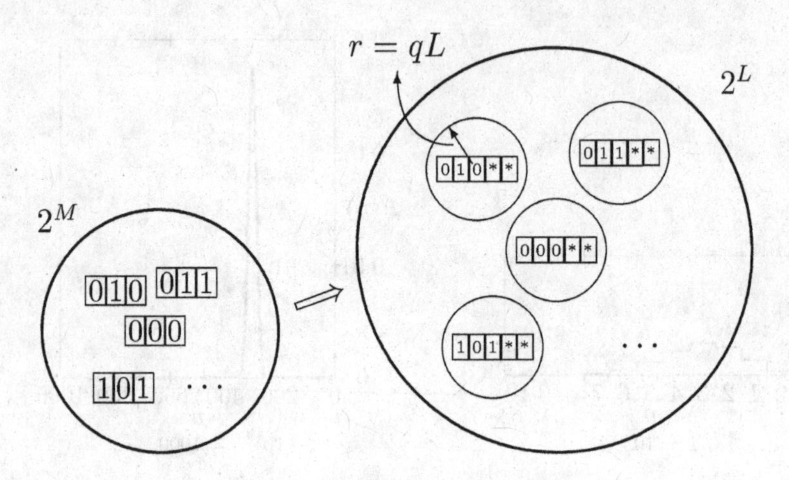


Fig. 7.4 Coding system from $M$-bit messages ($2^M$ combinations) to $L$-bit codes.

the original message from the wrong message. Therefore

$$2^M \leq \frac{2^L}{\binom{L}{qL}} \tag{7.17}$$

must hold. Assuming $L \gg 1$ and $Lq \gg 1$, let us apply Stirling's formula (7.5). We have

$$M \leq L\left[1 - q\log_2 q - (1-q)\log_2(1-q)\right]. \tag{7.18}$$

Thus the maximum amount of information is bound by

$$I \leq M/L \leq 1 - q\log_2\frac{1}{q} - (1-q)\log_2\frac{1}{1-q}. \tag{7.19}$$

Inversely, it is also proved that $I$ can reach to any close point of the upper limit at the right-hand side of Eq. (7.19).[1] One notes the limit is reached only with infinitely large $L$. and the coding system will become very complicated and lengthy. If the error rate is $q$=1/3, the upper limit on the efficiency is given by Eq.(7.19) as 0.08. That means we need 12 code bits for each data bit. In reality, more than this limit number is often used in the coding system.

[1] C. E. Shannon and W. Weaver, The Mathematical Theory of Communication (Univ. of Illinois Press, Chicago, 1949), which explains the first and the second theorems of Shannon in detail. One finds intuitive explanation in "Feynman Lectures on Computation" (Addison-Wesley, Reading, 1996).

**Summary of Chapter 7**

(1) The amount of information of an information source is given in terms of entropy. The lower bound of the size of the bit pattern to carry information is also given by the entropy (Source coding theorem, i.e., Shannon's first theorem).
(2) The capacity of a communication channel is bounded by its error rate (Channel coding theorem, i.e., Shannon's second theorem).

## 7.3 Problem

**[1]** Check the maximum and the minimum of $I$ defined by (7.2), where $0 \leq p_i \leq 1$ $(1 \leq i \leq N)$.

Chapter 8

# Quantum Computation

A mathematical model of present computers (classical computers) was invented by Alan Turing in 1930s, and it is called the Turing Machine. The procedure to solve problems on a classical computer is given as "instructions" on the Turing machine. The combination of these instructions is called "**algorithm**". Examples of the simplest algorithms are the arithmetic operations such as addition and subtraction. Among these algorithms, the multiplication is relatively fast, while those for division and chess game are rather slow. A typical one among slow algorithms is the factorization. It is an important subject for the model of computers to invent an effective algorithm for a certain problem. The effective algorithms mean that the instructions are finished in polynomial time. When one tries to solve the problem of factorization on a classical computer, the computational time increases exponentially, and it becomes practically impossible to solve the problem with a large number of digits. There is a problem called RSA139 in which the number to be solved has 139 digits. It took ten years to solve this RSA139 problem. The public key cryptography, one of the most advanced system of cryptography today, is based upon the difficulty of factorization.

In 1994, P. W. Shor proved that it is possible to perform the factorization of large numbers with remarkably fast speed (polynomial time) by using the quantum computer. By this discovery, the study on the quantum computer has entered a new era. In these years, theoretical and experimental studies of quantum computers spread widely all over the world.

**Keywords: Deutsch-Jozsa Algorithm; congruence; Shor's factorization algorithm; quantum phase estimation; order-finding algorithm.**

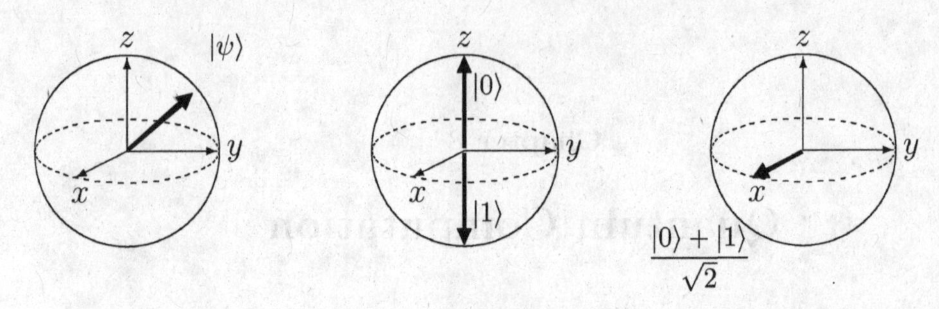


Fig. 8.1 Implementation of qubit by a superposition of quantum states. The middle figure shows two quantum states the spin-up and the spin-down states assigned $|0\rangle$ and $|1\rangle$, respectively. The right figure shows a qubit obtained by an Hadamard transformation (8.6).

## 8.1 Quantum bit and quantum register

The qubit (quantum bit) can be implemented by any quantum system with two states. For example, the spin-up $|\uparrow\rangle$ and the spin-down $|\downarrow\rangle$ states of electrons can be assigned to $|0\rangle$ and $|1\rangle$ (see Fig. 8.1). The set of qubits is called the quantum register, or simply the register. An integer 6 is expressed as 110 in the binary system. In the quantum register, this can be represented as a direct product of three states,

$$|6\rangle = |1\rangle \otimes |1\rangle \otimes |0\rangle. \tag{8.1}$$

The state (8.1) can be implemented by a three-bit register. In general, a number of n-bits

$$a = 2^{n-1}a_{n-1} + 2^{n-2}a_{n-2} + \cdots + 2^0 a_0. \tag{8.2}$$

can be expressed by a state of $n$-bit register,

$$\begin{aligned} |a\rangle &= |a_{n-1}\rangle \otimes |a_{n-2}\rangle \otimes \cdots \otimes |a_1\rangle \otimes |a_0\rangle \\ &\equiv |a_{n-1}a_{n-2}\cdots a_1 a_0\rangle. \end{aligned} \tag{8.3}$$

A quantum state of the $n$-bit register can be generalized to be a linear combination of states with numbers from $a = 0$ to $a = 2^n - 1$:

$$|\psi\rangle = \sum_{a=0}^{2^n-1} C_a |a\rangle. \tag{8.4}$$

where $C_a$ are complex numbers. The scalar product of the states $|a\rangle$ and $|b\rangle$ with $n$-bit register satisfies the orthonormalization condition

$$\begin{aligned} \langle a|b\rangle &= \langle a_0|b_0\rangle\langle a_1|b_1\rangle \cdots \langle a_{n-2}|b_{n-2}\rangle\langle a_{n-1}|b_{n-1}\rangle \\ &= \delta_{a,b}. \end{aligned} \tag{8.5}$$

Using the principles of quantum mechanics, we can construct new types of reversible (unitary) gates which do not exist in the classical gates. The Hadamard transformation $H$, for example, performs the operation,

$$H|0\rangle = \frac{1}{\sqrt{2}}(|0\rangle + |1\rangle) \tag{8.6}$$

$$H|1\rangle = \frac{1}{\sqrt{2}}(|0\rangle - |1\rangle). \tag{8.7}$$

This gate is implemented by the product of the spin-rotation operator $D_y^s\left(\frac{\pi}{2}\right)$ and the spin operator $\sigma_z$ in Eqs. (3.78) and (3.17), respectively;

$$H = D_y^s\left(\frac{\pi}{2}\right)\sigma_z = \left(\cos\left(\frac{\pi}{4}\right)\mathbf{1} - i\sin\left(\frac{\pi}{4}\right)\sigma_y\right)\sigma_z = \frac{1}{\sqrt{2}}\begin{pmatrix}1 & 1\\ 1 & -1\end{pmatrix}. \tag{8.8}$$

The rotation of spin can be achieved by the magnetic field such as the experimental set-up of Stern and Gerlach. Many quantum algorithms are based on so called **quantum parallelism**. As an effective algorithm by the quantum parallelism, we will show that the quantum computer can calculate the function $f(x)$ for many different values of $x$ by one instruction.

As a simple example, let us take the function $f(x)$,

$$x \in \{0,1\}, \quad f \in \{0,1\}, \tag{8.9}$$

where both $x$ and $f(x)$ have 1 bit size. We apply $f(x)$ to the state $|x,y\rangle$ with two registers $x$ and $y$. The first register $x$ is called the data register while the second register $y$ is called the target register. The operator $U_f$ acts on the state $|x,y\rangle$ as

$$U_f|x,y\rangle = |x, y \oplus f(x)\rangle, \tag{8.10}$$

where $y \oplus f(x)$ means the logical sum with mod 2. The operator $U_f$ is called **the Oracle operator** or **"the black box."**

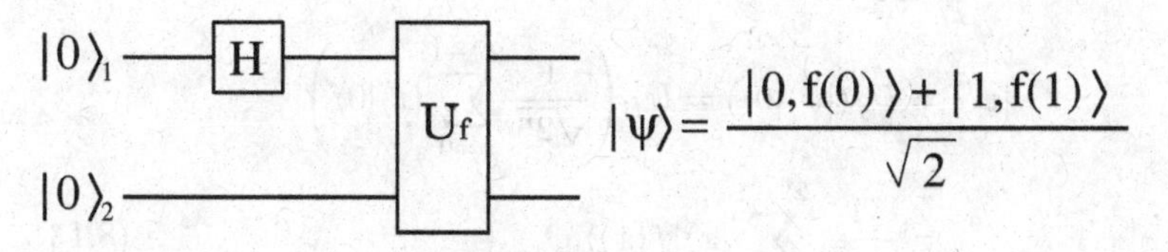


Fig. 8.2 Quantum parallelism.

As shown in Fig. 8.2, we apply first the Hadamard transformation to the state $|0\rangle$ of the data register. Secondly we apply $U_f$ to obtain the state

$$\begin{aligned}|\psi\rangle = U_f H|0,0\rangle &= U_f\frac{1}{\sqrt{2}}\{|0,0\rangle + |1,0\rangle\}\\ &= \frac{1}{\sqrt{2}}\{|0,f(0)\rangle + |1,f(1)\rangle\}.\end{aligned} \tag{8.11}$$

The state (8.11) arises as a result by the algorithm based on the quantum parallelism; the state $|\psi\rangle$ contains two results $f(0)$ and $f(1)$ as a linear combination. This is the fundamental difference from the parallelism in the classical computer. In the parallelism in the classical circuits, the different values of $f(x)$ are calculated in separate circuits. In the quantum parallelism, on the other hand, all the values of $f(x)$ are calculated in a single circuit. The concept of the quantum parallelism holds not only with a single bit but also with $n$ data bits. Let us write $H^{\otimes n}$ to express the Hadamard transformation acting on each bit of the $n$ bit state $|0\rangle^{\otimes n} = |00\cdots 0\rangle$. Then we have a linear combination of $2^n$ states:

$$\begin{aligned}|\psi\rangle = H^{\otimes n}|0\rangle &= H \otimes H \otimes \cdots \otimes H|00\cdots 0\rangle \\ &= \frac{1}{\sqrt{2}}(|0\rangle + |1\rangle) \otimes \frac{1}{\sqrt{2}}(|0\rangle + |1\rangle) \otimes \cdots \otimes \frac{1}{\sqrt{2}}(|0\rangle + |1\rangle) \\ &= \frac{1}{\sqrt{2^n}}(|00\cdots 0\rangle + |00\cdots 1\rangle + \cdots + |11\cdots 1\rangle) \\ &= \frac{1}{\sqrt{2^n}} \sum_{x=0}^{2^n-1} |x\rangle. \end{aligned} \tag{8.12}$$

The result (8.12) shows one of the most remarkable characteristics of quantum computation, i.e., it is possible to prepare the register with $2^n$ states (exponential function of $n$) by applying $H$ gate $n$ times (polynomial in $n$). This turns out to be an epoch-making algorithm that can never been achieved by classical gates.

Let us write the operation of the function $f(f : x \to f(x))$ on a quantum computer as $U_f$. By applying $U_f$ to state $|\psi\rangle$ of Eq. (8.12), we have

$$\begin{aligned}|\psi'\rangle = U_f(|\psi\rangle|0\rangle) &= U_f\left(\frac{1}{\sqrt{2^n}} \sum_{x=0}^{2^n-1} |x\rangle|0\rangle\right) \\ &= \frac{1}{\sqrt{2^n}} \sum_{x=0}^{2^n-1} |x\rangle|f(x)\rangle. \end{aligned} \tag{8.13}$$

The result of Eq. (8.13) shows that a single operation of $U_f$ produces a state which contains all of the $2^n$ values of the function $f(x)$. This conclusion is based on nothing but the principles of quantum mechanics. However when we consider the problem of quantum mechanical measurement, it is not very clear how effective the quantum parallelism is compared with the classical parallelism. In reality, the output register $f(x)$ is determined by the observation of value $x$ in the input register. This procedure implies that one output is determined by one input data just like the classical computer. Then, an intriguing question is what is the difference of this quantum algorithm to the classical algorithms? In Section 8.2, we will answer this question showing an example in which the quantum algorithm works in an effective way.

## 8.2 Deutsch-Jozsa algorithm

The effectiveness of the quantum parallelism was shown for the first time by Deutsch and Jozsa in 1992. The problem is the following special one.

---

**Example 8.1** (Deutsch-Jozsa problem): A function $f$ maps the set of $2^n$ integers

$$Z_{2^n} = \{0, 1, \ldots, 2^n - 1\} \tag{8.14}$$

onto two integers

$$Z_2 = \{0, 1\}. \tag{8.15}$$

Let us write this function $f$ as

$$f : Z_{2^n} \to Z_2. \tag{8.16}$$

Propose a quantum algorithm to find the truth among the following two propositions:

(a) The function $f$ is not constant.
(b) The number of zeros among $f(Z_i)$ ($i = 0, 1, \ldots, 2^n - 1$) is not $2^{n-1}$.

**Solution** It is easy to prove that at least one of the two propositions is true. If (a) is not true, then $f(Z_i)$ is a constant function, i.e., always 0 or always 1, namely, (b) is true. On the other hand, if (b) is not true, $f(Z_i)$ is not a constant function because it gives $2^{n-1}$ zeros and $2^{n-1}$ ones: (a) is true. In other cases, $f(Z_i)$ gives $n'$ ($n' \neq 2^{n-1}, n' \neq 2^n$) of 0 and $2^n - n'$ of 1: both (a) and (b) are true.

To solve this problem with the classical Turing machine, we need to call the subroutine $2^{n-1} + 1$ times to calculate $f(Z_i)$ $2^{n-1} + 1$ times. With the quantum computer, we will show that it is enough to call the subroutine three times.

We consider a quantum state

$$|Z_i, f(Z_i)\rangle \tag{8.17}$$

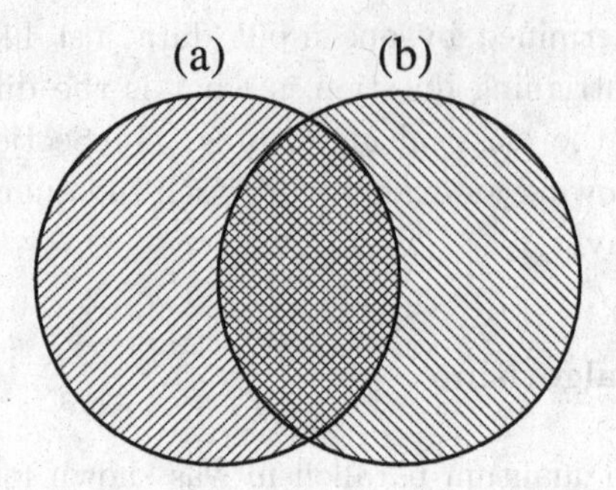


Fig. 8.3 The overlap of two circles shows the region where both (a) and (b) are true.

consisting of the input bit $Z_i$ and the corresponding output bit $f(Z_i)$. As the initial state, we take the linear combination of $2^n$ states,

$$|\psi_i\rangle = \frac{1}{\sqrt{2^n}}\left(\sum_{Z_i=0}^{2^n-1}|Z_i,0\rangle\right). \tag{8.18}$$

Let $U_f$ be a subroutine which calculates the function $f(Z_i)$ and gives its value into the output bit. By applying $U_f$ to the state (8.18), we obtain

$$|\psi_1\rangle = U_f|\psi_i\rangle = \frac{1}{\sqrt{2^n}}\left(\sum_{Z_i=0}^{2^n-1}|Z_i,f(Z_i)\rangle\right). \tag{8.19}$$

Then we apply the subroutine $\mathbf{1}\otimes\boldsymbol{\sigma}_z$ to calculate the phase of the output bits of the state $|\psi_1\rangle$ in Eq. (8.19);

$$|\psi_2\rangle = \mathbf{1}\otimes\boldsymbol{\sigma}_z|\psi_1\rangle = \frac{1}{\sqrt{2^n}}\sum_{Z_i=1}^{2^n-1}(-)^{f(Z_i)}|Z_i,f(Z_i)\rangle. \tag{8.20}$$

The inverse operation $U_f{}^{-1}$ of $U_f$ is applied on $|\psi_2\rangle$,

$$|\psi_f\rangle = \frac{1}{\sqrt{2^n}}\left(\sum_{Z_i=1}^{2^n-1}(-)^{f(Z_i)}|Z_i,0\rangle\right). \tag{8.21}$$

Equation (8.21) is the final state which will provide the solution. The overlap between the initial and the final states is given by

$$P\equiv|\langle\psi_f|\psi_i\rangle|^2 = \left(\frac{1}{2^n}\right)^2\left|\sum_{Z_i=0}^{2^n-1}(-)^{f(Z_i)}\right|^2. \tag{8.22}$$

By measuring the value of (8.22), we can discern the truth or the falsehood from the provisions as

$$P = \begin{cases} 1 & \text{(b) is true and (a) is false;} \\ 0 & \text{(a) is true and (b) is false;} \\ 0 < P < 1 & \text{both (a) and (b) are true.} \end{cases} \tag{8.23}$$

As we have seen, the provision needs to call the subroutine $2^{n-1} + 1$ times in the classical computer. Surprisingly, this problem can be treated in the parallel algorithm of quantum computation by calling the three subroutines, $U_f$, $U_f^{-1}$ and the evaluation of phase.

---

The Deutsch-Jozsa algorithm is very interesting showing the effectiveness of quantum computers, but the problem is a very conceptual one which does not lead to any practical application. In the next section, we will see the quantum algorithm presented by Shor which has possibilities of important applications.

## 8.3 Shor's factorization algorithm

Let us consider a problem of factoring an integer of 20 digits,

$$N = 39772916239307209103. \tag{8.24}$$

How can we solve the problem? If $N$ has factors $p, q$, i.e., $N = pq$, one of them, $p$ or $q$ should be smaller than $\sqrt{N}$ or equal to $\sqrt{N}$. The most primitive way to find prime factors will be to divide $N$ by all the integers starting from 1 up to $\sqrt{N}$. In the case of (8.24), we will need billions of division because $\sqrt{N}$ has ten digits. This algorithm needs to call the subroutine by

$$\sqrt{N} = 2^{\frac{1}{2}\log_2 N} \tag{8.25}$$

times, i.e., the amount of computation increases as an exponential function of the number of digits of $N$. Therefore this is not an effective algorithm. On the other hand, the problem of multiplying two integers:

$$\begin{aligned} p &= 6257493337 \\ q &= 6356046119 \end{aligned} \tag{8.26}$$

is solved by calling the subroutine only once so that this is an effective algorithm in polynomial time. To find effective algorithms for the factorization of big numbers has a close connection with the theory of cryptography, and becomes one of the most important problems of algorithm in the modern time.

In 1994, Shor showed that the problem of factorization can be solved in polynomial time by using the quantum parallel algorithm. Shor's algorithm is based on the fact that the factorization of large integers can be attributed to the problem of finding the period of **the congruence equation** modulo $N$. Making use of the number theory, the problem of finding prime factors $p$ and $q$ of a given integer $N$ can be converted to the problem of finding the period of the function $f(a)$ defined by the following congruence equation for a given $x$:

$$f(a) \equiv x^a \bmod N. \tag{8.27}$$

where $x$ is an integer prime to $N$, and is less than $N$. The modular exponential $x^a \bmod N$ gives the residue modulo $N$. Here "congruence" has the same meaning as in geometry based on the following idea: $f(a)$ and $x^a$ is equal as long as the difference in a multiple of $N$ is ignored. The same concept is given in geometry that two objects are equal as long as the difference in location is ignored. The function $f(a)$ in (8.27) is periodic. In particular, in the case of $f(a) = 1$, it is expressed as

$$x^r \equiv 1 \bmod N, \tag{8.28}$$

and $r$ is called **the order** of $x$ in the congruence modulo $N$. If $r$ is the order of $x \bmod N$ and is an even number, we have

$$x^r - 1 = (x^{\frac{r}{2}} - 1)(x^{\frac{r}{2}} + 1) = nN \ \ (n : \text{integer}), \tag{8.29}$$

and the greatest common divisor (gcd) of $x^{\frac{r}{2}} \pm 1$ and $N$,

$$\gcd(x^{\frac{r}{2}} + 1, N), \ \ \gcd(x^{\frac{r}{2}} - 1, N) \tag{8.30}$$

will be a factor of $N$ with high probability. Though $r = 0$ gives an apparent solution of $x^r = 1 \bmod N$, $r \neq 0$ is the one which we want. Noting that $f(a)$ is a periodic function, we see that the order $r$ coincides with the period of $f(a)$. Therefore, for the purpose of finding the order $r$, it is enough to find the period of $f(a)$. That is the basic idea of factorization by Shor's quantum computation. Now we show a specific example of the application of Shor's algorithm.

---

**Example 8.2** Factorize $N = 15$ on a quantum computer.

**Solution** The procedure of factorization on a quantum computer is as follows.

**Step 1**:

Take an integer $x$ that is less than, and prime to $N$. Here we choose $x = 7$.

**Step 2**:

In order to find the order $r$ of the congruence

$$x^r \equiv 1 \bmod N, \tag{8.31}$$

we perform the Hadamard transformation $n$ times on the first register of the initial state $|0\rangle^{\otimes n}|1\rangle$ given by (8.8):

$$H^{\otimes n}|0\rangle^{\otimes n}|1\rangle = \frac{1}{\sqrt{2^n}}\sum_{a=0}^{2^n-1}|a\rangle|1\rangle \tag{8.32}$$

Let us adopt $n = 11$.

**Step 3**:

We calculate $f(a) \equiv x^a \bmod N$, and write the result on the second register:

$$\frac{1}{\sqrt{2^n}}\sum_{a=0}^{2^n-1}|a\rangle|x^a \bmod N\rangle$$
$$= \frac{1}{\sqrt{2^n}}[|0\rangle|1\rangle + |1\rangle|7\rangle + |2\rangle|4\rangle + |3\rangle|13\rangle + |4\rangle|1\rangle + |5\rangle|7\rangle + |6\rangle|4\rangle + \cdots]. \tag{8.33}$$

**Step 4**:

By measuring the second register, we find that $r = 0, 4, 8, \ldots$ satisfy $x^r \equiv 1 \bmod N$.

**Step 5**:

From $x^{\frac{r}{2}} \pm 1 = 7^2 \pm 1 = 48, 50$, we have $\gcd(48, 15) = 3$ and $\gcd(50, 15) = 5$. Thus we obtain 3 and 5 as the prime factors of $N = 15$.

---

The quantum computer for the factorization of an integer $N$ of $L$ bits is designed in the following way. First, you prepare the first register with $n = 2L + 1$ bits and the second register with $L$ bits: $|0\rangle|1\rangle$. Take an integer $x$ ($\gcd(N, x) = 1$) which is less than, and prime to $N$. The reason for choosing $n = 2L + 1$ bits for the first register is to make the probability of finding the order $r$ large when it is prime to $q = 2^n$. The summary of Shor's algorithm is summarized as:

**Step 1**:

Prepare the state $|\psi_i\rangle = |0\rangle|1\rangle$ with the first register with $n = 2L + 1$ bits and the second register with $L$ bits.

**Step 2**:

Perform the Hadamard transformation on the first register by $n$ times, where $q = 2^n$;

$$|\psi_2\rangle = \sum_{a=0}^{q-1} |a\rangle|1\rangle \tag{8.34}$$

**Step 3**:

(Order-finding algorithm) Calculate $x^a$ mod $N$ and store it on the second register.

$$|\psi_3\rangle = \sum_{a=0}^{q-1} |a\rangle|x^a \bmod N\rangle \tag{8.35}$$

**Step 4**:

Measure the second register and select the numbers $k$ from the first register , which satisfy the periodicity $x^k$ mod $N = 1$.

**Step 5**:

Perform the inverse Fourier transformation on the first register and measure the probabilities to obtain the order $r$ from their periodicity.

## 8.4 Quantum Fourier transformation with $n$ bits

The order-finding and the quantum Fourier transformation (QFT) play important roles in the circuit of the quantum computer implementing Shor's factorization algorithm. In this section, we study how to perform the quantum Fourier transformation of $n$ bits. The order-finding algorithm will be discussed in Section 8.5. The Fourier transformation is a unitary transformation with the basis of $q = 2^n$ dimensions,

$$|a\rangle \qquad (a = 0, \ldots, q-1) \tag{8.36}$$

and gives the new state

$$\mathrm{QFT}_q : |a\rangle \longrightarrow \frac{1}{\sqrt{q}} \sum_{c=0}^{q-1} \exp(2\pi iac/q)|c\rangle. \tag{8.37}$$

For the state $|0\rangle$, QFT gives a linear combination of all the basis states with equal amplitudes. The Fourier transformation in the $q = 2^n$-dimensional space is implemented by the combination of two quantum gates. One is the $2 \times 2$ Hadamard gate given by Eq. (8.8),

$$H_j = \frac{1}{\sqrt{2}} \begin{pmatrix} 1 & 1 \\ 1 & -1 \end{pmatrix} \begin{matrix} |0\rangle \\ |1\rangle \end{matrix}. \quad \text{(columns: } |0\rangle, |1\rangle\text{)} \tag{8.38}$$

operating on the $j$th bit. The other is the controlled phase gate, i.e., the two-quantum-bit gates operating on the $k$th and the $j$th bits $(k > j)$

$$U_{jk}(\theta_{jk}) = \begin{pmatrix} 1 & 0 & 0 & 0 \\ 0 & 1 & 0 & 0 \\ 0 & 0 & 1 & 0 \\ 0 & 0 & 0 & e^{i\theta_{jk}} \end{pmatrix} \begin{matrix} |00\rangle \\ |01\rangle \\ |10\rangle \\ |11\rangle \end{matrix}, \quad \text{(columns: } |00\rangle, |01\rangle, |10\rangle, |11\rangle\text{)} \tag{8.39}$$

where $\theta_{jk} \equiv \frac{\pi}{2^{j-k}}$. The control gate $U_{jk}$ operates on two bits and changes the phase of the target bit $k$ by $e^{i\theta_{jk}}$ only if the two bits $|a_j, a_k\rangle$ satisfy the condition $a_j = a_k = 1$. Let us design an QFT circuit using the transformations $H_j$ and $U_{jk}$.

Let us consider the $q$-dimensional quantum Fourier transformation

$$\mathrm{QFT}_q|a\rangle = \frac{1}{\sqrt{q}} \sum_{c=0}^{q-1} e^{2\pi iac/q}|c\rangle \tag{8.40}$$

where

$$|a\rangle = |a_{n-1}, a_{n-2}, a_{n-3}, \ldots, a_1, a_0\rangle \tag{8.41}$$

$$|c\rangle = |c_{n-1}, c_{n-2}, c_{n-3}, \ldots, c_1, c_0\rangle. \tag{8.42}$$

The numbers $a$ and $c$ are expressed in the binary expression as

$$a = 2^{n-1}a_{n-1} + 2^{n-2}a_{n-2} + \cdots + 2^2a_2 + 2^1a_1 + 2^0a_0 \tag{8.43}$$

$$c = 2^{n-1}c_{n-1} + 2^{n-2}c_{n-2} + \cdots + 2^2c_2 + 2^1c_1 + 2^0c_0. \tag{8.44}$$

In Chapter 6, we have treated the example with $q = 2^3$. The same algorithm can be generalized to the $q = 2^n$-dimensional case as

$$
\begin{aligned}
&\mathrm{QFT}_q|a\rangle \\
&= \mathrm{QFT}_q|a_{n-1}, a_{n-2}, \ldots, a_1, a_0\rangle \\
&= \frac{1}{\sqrt{2^n}} \sum_{c_{n-1}=0}^{1} \sum_{c_{n-2}=0}^{1} \cdots \sum_{c_1=0}^{1} \sum_{c_0=0}^{1} e^{2\pi i(a_0+2a_1+\cdots+2^{n-2}a_{n-2}+2^{n-1}a_{n-1})\times} \\
&\qquad {}^{\times(c_0+2c_1+\cdots+2^{n-2}c_{n-2}+2^{n-1}c_{n-1})/2^n} |c_{n-1}, c_{n-2}, \ldots, c_2, c_1, c_0\rangle \\
&= \left(\frac{1}{\sqrt{2}} \sum_{c_{n-1}=0}^{1} e^{2\pi i 2^{n-1} c_{n-1}(a_0+2a_1+\cdots+2^{n-2}a_{n-2}+2^{n-1}a_{n-1})/2^n} |c_{n-1}\rangle_{n-1}\right) \\
&\otimes \left(\frac{1}{\sqrt{2}} \sum_{c_{n-2}=0}^{1} e^{2\pi i 2^{n-2} c_{n-2}(a_0+2a_1+\cdots+2^{n-2}a_{n-2}+2^{n-1}a_{n-1})/2^n} |c_{n-2}\rangle_{n-2}\right) \\
&\otimes \cdots\cdots\cdots \\
&\vdots \\
&\otimes \left(\frac{1}{\sqrt{2}} \sum_{c_1=0}^{1} e^{2\pi i 2 c_1(a_0+2a_1+\cdots+2^{n-2}a_{n-2}+2^{n-1}a_{n-1})/2^n} |c_1\rangle_1\right) \\
&\otimes \left(\frac{1}{\sqrt{2}} \sum_{c_0=0}^{1} e^{2\pi i c_0(a_0+2a_1+\cdots+2^{n-2}a_{n-2}+2^{n-1}a_{n-1})/2^n} |c_0\rangle_0\right) \\
&= \frac{1}{\sqrt{2}} \left(|0\rangle_{n-1} + e^{2\pi i a_0/2} |1\rangle_{n-1}\right) \\
&\otimes \frac{1}{\sqrt{2}} \left(|0\rangle_{n-2} + e^{2\pi i\left(a_1/2+a_0/2^2\right)} |1\rangle_{n-2}\right) \\
&\otimes \cdots\cdots\cdots \\
&\vdots \\
&\otimes \frac{1}{\sqrt{2}} \left(|0\rangle_1 + e^{2\pi i\left(a_{n-2}/2+a_{n-3}/2^2+\cdots+a_1/2^{n-2}+a_0/2^{n-1}\right)} |1\rangle_1\right) \\
&\otimes \frac{1}{\sqrt{2}} \left(|0\rangle_0 + e^{2\pi i\left(a_{n-1}/2+a_{n-2}/2^2+\cdots+a_1/2^{n-1}+a_0/2^n\right)} |1\rangle_0\right). \qquad (8.45)
\end{aligned}
$$

The $n$-bit QFT gate is shown in Fig. 8.4 as a generalized form of Fig. 6.17. The inverse quantum discrete transformation can be constructed by the reversed time evolution of the operations in Fig. 8.4.

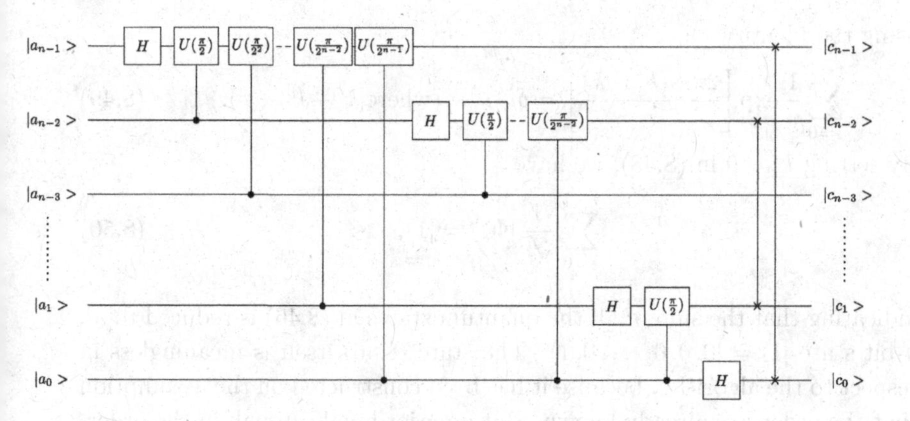


Fig. 8.4 QFT gate with $n$ bits.

## 8.5 Quantum phase estimation and order finding algorithm

One of the important algorithms in Shor's factorization is the order-finding in the congruence equation. On classical computers, no effective algorithm has been discovered so far. In Shor's algorithm in Section 8.3, the operation of congruence equations for the order-finding was performed on the second register, i.e., the period of congruence equations was measured to find the order. In the present section, we will see that the order-finding can be attained by making use of the quantum superposition without measuring the second register . This is a typical example where the principle of quantum superposition provides the effective algorithm in a remarkable way.

To begin with, let us consider a quantum state

$$|\Phi_s\rangle = \frac{1}{\sqrt{r}} \sum_{k=0}^{r-1} \exp\left[\frac{-2\pi isk}{r}\right] |x^k \bmod N\rangle, \tag{8.46}$$

assuming that the order $r$ is known. In Eq. (8.46), $s$ is an integer in the range: $0 \leq s \leq r-1$. Since the state $|\Phi_s\rangle$ is a quantum Fourier transformed state, the inverse Fourier transform of (8.46),

$$\sum_{s=0}^{r-1} \frac{1}{\sqrt{r}} \exp\left[\frac{2\pi isk'}{r}\right], |\Phi_s\rangle \tag{8.47}$$

is converted into

$$\sum_{s=0}^{r-1} \frac{1}{\sqrt{r}} \exp\left[\frac{2\pi isk'}{r}\right] |\Phi_s\rangle = |x^{k'} \bmod N\rangle, \tag{8.48}$$

using the identity

$$\sum_{s=0}^{r-1} \frac{1}{r} \exp\left[\frac{2\pi i s(k'-k)}{r}\right] = \delta_{k',k} \quad (\text{where} |k'-k| < r). \tag{8.49}$$

By setting $k' = 0$ in (8.48), we have

$$\sum_{s=0}^{r-1} \frac{1}{\sqrt{r}} |\Phi_s\rangle = |1\rangle, \tag{8.50}$$

indicating that the sum of all the quantum states in (8.46) is reduced to an $n$-bit state $|1\rangle = |0, 0, 0, \ldots, 0, 1\rangle$. The state (8.46) itself is meaningless in respect to the algorithm because it has been constructed on the assumption that the order $r$ is already known. But a major breakthrough in the order-finding algorithm is hidden in the fact that the sum (8.50) is identified as the $n-$bit state $|1\rangle$.

Let us choose integers $y$ and $N$ which satisfy $0 \leq y \leq N-1$ ($N < 2^n$). Then define a unitary operator $U_x$

$$U_x|y\rangle \equiv |xy \bmod N\rangle, \tag{8.51}$$

where $x$ is another integer prime to $N$, i.e., $\gcd(x, N) = 1$. Applying this operator to the state (8.46), we have

$$\begin{aligned} U_x|\Phi_s\rangle &= \frac{1}{\sqrt{r}} \sum_{k=0}^{r-1} \exp\left[-\frac{2\pi i s k}{r}\right] |x^{k+1} \bmod N\rangle & (8.52) \\ &= \exp\left[\frac{2\pi i s}{r}\right] |\Phi_s\rangle. & (8.53) \end{aligned}$$

In this way, it has been shown that the state $|\Phi_s\rangle$ is the eigenstate of $U_x$ with the eigenvalue $\exp[2\pi i s/r]$. In other words, when $U_x$ operate on $|\Phi_s\rangle$, it works as a phase operator giving a phase $\exp[2\pi i s/r]$.

---

**Example 8.3** Prove (8.53).

**Solution** By setting $k' = k+1$ in Eq. (8.52), we have

$$U_x|\Phi_s\rangle = \frac{1}{\sqrt{r}} \exp\left[\frac{2\pi i s}{r}\right] \sum_{k'=1}^{r} \exp\left[-\frac{2\pi i s k'}{r}\right] |x^{k'} \bmod N\rangle. \tag{8.54}$$

In order to prove Eq. (8.53), it is enough to show the term with $k' = r$ has the same contribution as the term with $k' = 0$. This can be shown by

$$\exp\left[-\frac{2\pi i s r}{r}\right] |x^r \bmod N\rangle = |x^r \bmod N\rangle = |1\rangle = |x^0 \bmod N\rangle. \tag{8.55}$$

In this way, we have proved that Eq. (8.53) holds, i.e., $|\Phi_s\rangle$ is an eigenstate of operator $U$.

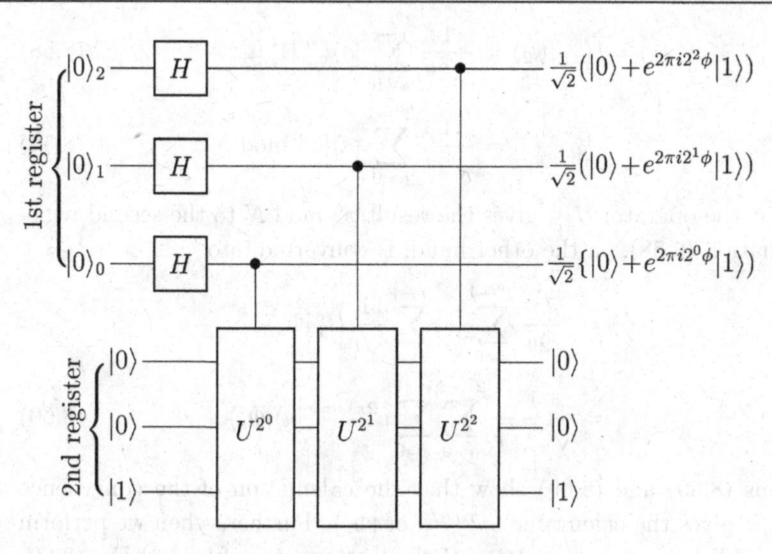


Fig. 8.5 Phase Computation with a 3-bit Quantum Circuit. This is the circus how to implement the phase computation which is needed for the unitary operator $U_x^a$ in Eq. (8.58).

The application of $U_x$ by "$a$" times gives rise to the operator $U_x^a$ which maps $|y\rangle$ onto $|x^a y\rangle$. We further extend $U_x^a$ and define the operator $U_{x,a}$ which operates on two registers:

$$U_{x,a}|a\rangle|y\rangle \equiv |a\rangle U_x^a|y\rangle = |a\rangle|x^a y \bmod N\rangle. \tag{8.56}$$

Let us think about the order-finding algorithm with this operator $U_{x,a}$. As shown in Fig. 8.5, we prepare the initial state: $|\psi_1\rangle = |0\rangle^{\otimes n}|1\rangle$. Let $L = \lceil \log_2 N \rceil$ ($\lceil x \rceil$ means the smallest integer larger than $x$) and prepare the first register with $n = 2L + 1$ bits and the second register with $L$ bits. The number of bits on the first register has been chosen for the sake of efficient search of the order $r$. First, we perform $n$-bit Hadamard transformation on the first register,

$$|\psi_2\rangle = H^{\otimes n}|\psi_1\rangle = \frac{1}{\sqrt{2^n}} \sum_{a=0}^{2^n-1} |a\rangle|1\rangle, \tag{8.57}$$

where the symbol $|1\rangle$ of the second register stands for the state $|1\rangle$ of $L$ bits. Next we perform the unitary operator $U_{x,a}$ to the state (8.57),

$$|\psi_3\rangle = U_{x,a}|\psi_2\rangle = \frac{1}{\sqrt{2^n}} \sum_{a=0}^{2^n-1} |a\rangle U_x^a |1\rangle \tag{8.58}$$

$$= \frac{1}{\sqrt{2^n}} \sum_{a=0}^{2^n-1} |a\rangle |x^a \bmod N\rangle. \tag{8.59}$$

Note that the operator $U_{x,a}$ gives the result $x^a$ mod $N$ to the second register. Equation (8.58), on the other hand, is converted into

$$|\psi_3'\rangle = \frac{1}{\sqrt{2^n}} \sum_{a=0}^{2^n-1} |a\rangle \sum_{s=0}^{r-1} \frac{1}{\sqrt{r}} U_x^a |\Phi_s\rangle$$

$$= \frac{1}{\sqrt{2^n \cdot r}} \sum_{a=0}^{2^n-1} \sum_{s=0}^{r-1} e^{2\pi i s a/r} |a\rangle |\Phi_s\rangle. \tag{8.60}$$

Equations (8.59) and (8.60) show that the calculation of the congruence $x^a$ mod $N$ gives the eigenvalue $e^{2\pi i s a/r}$ of $|\Phi_s\rangle$. Further, when we perform the inverse Fourier transformation of the first register of $|\psi_3'\rangle$ of Eq. (8.60) and sum over $a$, we have

$$|\psi_4\rangle = QFT^\dagger |\psi_3'\rangle$$

$$= \frac{1}{2^n} \sum_{a,c=0}^{2^n-1} \frac{1}{\sqrt{r}} \sum_{s=0}^{r-1} e^{-2\pi i c a/2^n} e^{2\pi i s a/r} |c\rangle |\Phi_s\rangle$$

$$= \frac{1}{2^n} \sum_{a,c=0}^{2^n-1} \frac{1}{\sqrt{r}} \sum_{s=0}^{r-1} e^{2\pi i (s/r - c/2^n) a} |c\rangle |\Phi_s\rangle. \tag{8.61}$$

If $2^n$ is divisible by the order $r$, we have

$$|\psi_4\rangle = \frac{1}{\sqrt{r}} \sum_{s=0}^{r-1} \sum_{c=0}^{2^n-1} \delta_{c, s \cdot 2^n/r} |c\rangle |\Phi_s\rangle$$

$$= \frac{1}{\sqrt{r}} \sum_{s=0}^{r-1} \left| s \frac{2^n}{r} \right\rangle |\Phi_s\rangle \tag{8.62}$$

using the identity (8.49). Equation (8.62) shows that the order $r$ can be found by measuring the period of the first register. Though the existence of the eigenstates in the second register $\Phi_s$ has been introduced only by assumption, the order-finding can be performed without the observation of the second register.

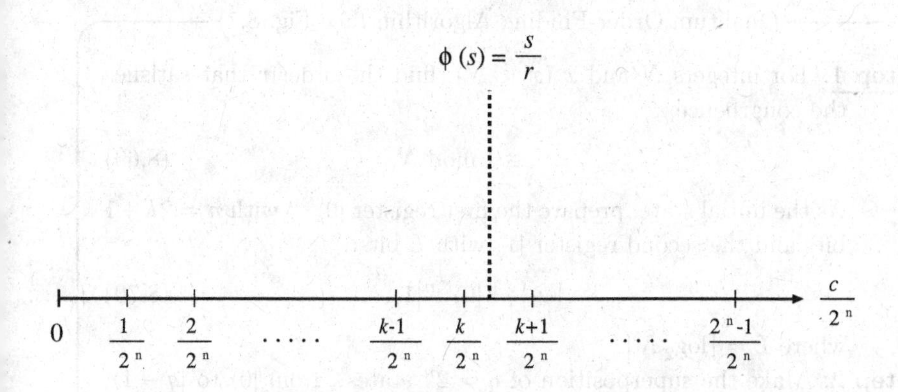


Fig. 8.6 Order-finding in the case where $2^n$ is indivisible by $r$.

Next we consider the case where $2^n$ is indivisible by $r$. Assuming that the order $r$ was found a priori, let $\delta$ be the difference between $\phi(s) = s/r$ and $c/2^n$:

$$\phi(s) = \frac{c}{2^n} + \delta. \tag{8.63}$$

As seen in Fig. 8.6, there must be $c$ in the range $c \in (0, 1, \cdots, 2^n - 1)$.

$$|\delta| \leq \frac{1}{2 \cdot 2^n}. \tag{8.64}$$

Let us denote that value of $c$ as $k$. Then we have

$$\left|\phi(s) - \frac{k}{2^n}\right| \leq \frac{1}{2 \cdot 2^n}. \tag{8.65}$$

From the continued fraction theorem (A.87) in the Appendix, $\frac{k}{2^n}$ is an approximation to $\phi(s)$ in the form of continued fraction. Incidentally, one notes that if Eq. (8.65) is satisfied, the probability of the state $|k\rangle$ is given by

$$P(k) = \frac{1}{r}\frac{1}{2^{2n}}\left|\sum_{a=0}^{2^n-1} e^{2\pi i a(\phi(s)-k/2^n)}\right|^2. \tag{8.66}$$

If $0 \leq \phi(s) - \frac{k}{2^n} \leq \frac{1}{2 \cdot 2^n}$ in Eq. (8.65), the phase of Eq. (8.66) becomes

$$0 \leq 2\pi a\left(\phi(s) - \frac{k}{2^n}\right) \leq \pi\frac{a}{2^n} < \pi. \tag{8.67}$$

If $-\frac{1}{2 \cdot 2^n} \leq \phi(s) - \frac{k}{2^n} < 0$, the phase becomes

$$-\pi < -\pi\frac{a}{2^n} \leq 2\pi a\left(\phi(s) - \frac{k}{2^n}\right) \leq 0. \tag{8.68}$$

Quantum Order-Finding Algorithm (See Fig. 8.7)

**Step 1**: For integers $N$ and $x$ ($x < N$), find the order $r$ that satisfies the congruence

$$x^r \equiv 1 \bmod N. \tag{8.69}$$

As the initial state, prepare the first register $|0\rangle^{\otimes n}$ with $n = 2L+1$ bits and the second register $|1\rangle$ with $L$ bits:

$$|\psi_1\rangle = |0\rangle^{\otimes n}|1\rangle, \tag{8.70}$$

where $L = \lceil \log_2 N \rceil$.

**Step 2**: Make the superposition of $q = 2^n$ states , from $|0\rangle$ to $|q-1\rangle$, by the Hadamard transformation $H^{\otimes n}$ on the first register,

$$|\psi_2\rangle = \frac{1}{\sqrt{2^n}} \sum_{a=0}^{q-1} |a\rangle|1\rangle \tag{8.71}$$

**Step 3**: Apply $U_{x,a}$ to $|\psi_2\rangle$:

$$\begin{aligned}|\psi_3\rangle = U_{x,a}|\psi_2\rangle &= \frac{1}{\sqrt{2^n}} \sum_{a=0}^{q-1} |a\rangle U_x^a|1\rangle \\ &= \frac{1}{\sqrt{2^n}} \sum_{a=0}^{q-1} |a\rangle \frac{1}{\sqrt{r}} U_x^a |\Phi_s\rangle = \frac{1}{\sqrt{2^n r}} \sum_{a=0}^{q-1} \sum_{s=0}^{r-1} e^{2\pi i s a/r} |a\rangle|\Phi_s\rangle.\end{aligned} \tag{8.72}$$

**Step 4**: Perform the inverse Fourier transformation on the first register:

$$\begin{aligned}|\psi_4\rangle = \mathrm{QFT}^\dagger|\psi_3\rangle &= \frac{1}{2^n} \sum_{a,c=0}^{q-1} \frac{1}{\sqrt{r}} \sum_{s=0}^{r-1} e^{-2\pi i c a/q} e^{2\pi i s a/r} |c\rangle|\Phi_s\rangle \\ &= \frac{1}{\sqrt{r}} \sum_{s=0}^{r-1} \frac{1}{2^n} \sum_{a,c=0}^{q-1} e^{2\pi i (s/r - c/q)a} |c\rangle|\Phi_s\rangle \\ &= \frac{1}{\sqrt{r}} \sum_{s=0}^{r-1} \left| \widetilde{s\frac{2^n}{r}} \right\rangle |\Phi_s\rangle.\end{aligned} \tag{8.73}$$

Note that there are two possibilities with the state $|\widetilde{s\frac{2^n}{r}}\rangle$: $2^n$ is divisible or indivisible by $r$.

**Step 5**: Measure the probability of the first register. If the probability shows the periodicity of $s/r$, it is the order $r$.

Thus if the condition (8.65) is fulfilled, all the terms of the sum (8.66) spread only in the upper (lower) part of the complex plane, and the sum over $a$ has positive interference and becomes finite. If (8.65) is not satisfied, the terms in the sum spread over the entire complex plane and cancel one another. As a result, the sum becomes negligibly small. In this way, we can find the order $r$ with large probability even when $r$ does not divide $2^n$.

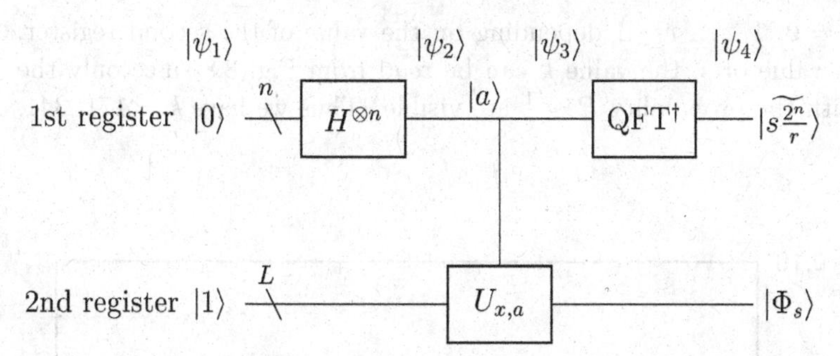


Fig. 8.7 Quantum circuit for order-finding. The first register has the quantum state with $n = 2L + 1$ bits. The second register has the initial state $|1\rangle$ with $L$ bits. The symbol $H^{\otimes n}$ means the multiple Hadamard transformation by $n$ times. The symbol QFT$^\dagger$ stands for the inverse Fourier transformation.

---

**Example 8.4** Factorize $N = 35$ by quantum algorithm

**Solution** Take $L = \log_2 35 \approx 5$, $n = 2L + 1 = 11$ and $q = 2^{11} = 2048$. As an integer prime to $N$, let us choose $x = 3$. First, prepare the state

$$\frac{1}{\sqrt{q}}\sum_{a=0}^{q-1}|a\rangle|1\rangle = \frac{1}{\sqrt{q}}[|0\rangle + |1\rangle + |2\rangle + \cdots + |q-1\rangle]|1\rangle. \tag{8.74}$$

by performing the Hadamard transformation. Next, we evaluate the modular exponential $x^a$ mod $N$ by calling the order-finding subroutine $U_{x,a}$, and store it in the second register:

$$\begin{aligned}&\frac{1}{\sqrt{q}}\sum_{a=0}^{q-1}|a\rangle|x^a \bmod N\rangle \\ &= \frac{1}{\sqrt{q}}[|0\rangle|1\rangle + |1\rangle|3\rangle + |2\rangle|9\rangle + |3\rangle|27\rangle + |4\rangle|11\rangle + |5\rangle|33\rangle \\ &\quad +|6\rangle|29\rangle + |7\rangle|17\rangle + |8\rangle|16\rangle + |9\rangle|13\rangle + |10\rangle|4\rangle + |11\rangle|12\rangle \\ &\quad +|12\rangle|1\rangle + |13\rangle|3\rangle + |14\rangle|9\rangle + \cdots].\end{aligned} \tag{8.75}$$

As a result, we see twelve integers, 1, 3, 9, 27, 11, 33, 29, 17, 16, 13, 4, 12 in the second register in the apparently random order. Let us choose one of them, perform the inverse Fourier transformation, and measure the value $c$ of the first register. The result is given by the probability:

$$P(c) = \frac{1}{r}\frac{1}{q^2}\left|\sum_{a=0}^{q-1} e^{2\pi i(\phi(s)-c/2^n)a}\right|^2, \tag{8.76}$$

where $s = 0, 1, \ldots, r-1$ depending on the value of the second register. For each value of $s$, the value $k$ can be read from Fig. 8.8 since only the states with the probability $P \sim \frac{1}{r}$ are visible. Thus we have $k = 171, 341,$

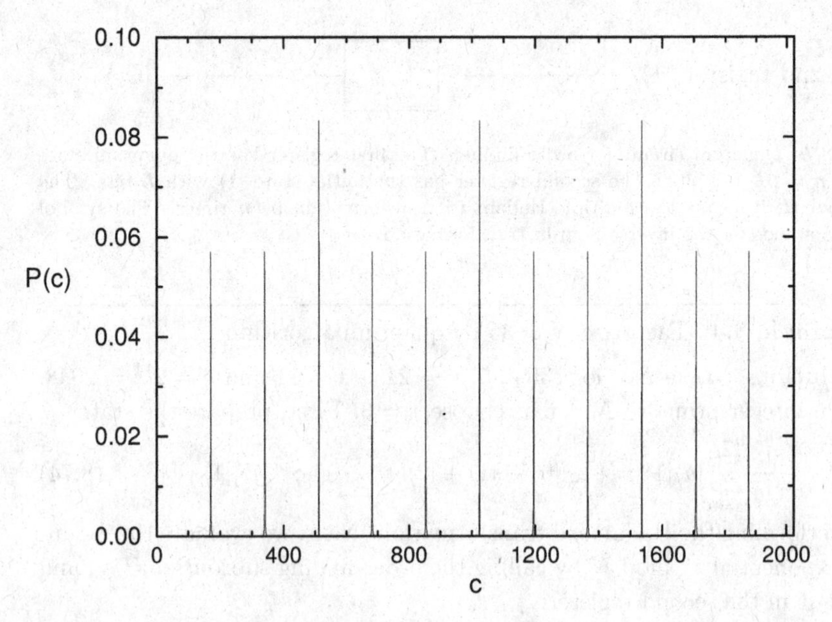


Fig. 8.8 The probability of observing $|c\rangle$ in the first register after the quantum Fourier transformation, where $q = 2^{11} = 2048$.

512, 683, 853, 1024, 1195, 1365, 1536, 1707, 1877. Choosing $k = 853$, we expand $k/r = 853/2048$ in terms of a continued fraction and obtain

$$\frac{853}{2048} \cong \cfrac{1}{2+\cfrac{1}{2+\cfrac{1}{2}}} = \frac{5}{12}$$

as the value of $s/r$ satisfying (8.65). If we choose $k = 1195$ instead, we have

$$\frac{1195}{2048} \cong \cfrac{1}{1+\cfrac{1}{1+\cfrac{1}{2+\cfrac{1}{2}}}} = \frac{7}{12}$$

as the converged value. In either case, we find the order $r = 12$. For $r = 12$, we can obtain

$$x^{r/2} \pm 1 = 3^6 \pm 1 = 728,\ 730$$

and

$$\gcd(728, 35) = 7, \qquad \gcd(730, 35) = 5.$$

Finally the factorization is achieved as $35 = 7 \times 5$. Note that if we had chosen $k = 1024$, we would have $\frac{k}{r} = \frac{1}{2}$. This case the factorization is unsuccessful.

---

The factorization by Shor's algorithm was experimentally implemented using the NMR quantum computer with the molecule qubits consisting of seven atomic nuclei with spin $\frac{1}{2}$. The experiment for factorization of $N = 15$ was successfully completed and published in December, 2001 (See Section 11.1.4). It is shown that the discrete logarithm problem which is intractable by classical computers can be solvable also by a quantum computer making use of the Oracle $U$ which is used in the calculations of congruence equations.

## 8.6 Calculation of modular exponentials

In the order-finding algorithm, we adopt the control operation $U_{x,a}$. Now let us think about how to implement $U_{x,a}$ as a quantum circuit . Applying $U_{x,a}$ to the states on the right-hand side of (8.57), we have

$$|a\rangle|1\rangle \xrightarrow{U_{x,a}} |a\rangle U_x^a|1\rangle = |a\rangle|x^a \bmod N\rangle. \tag{8.77}$$

Expressing $a$ as $a = a_{n-1}2^{n-1} + a_{n-2}2^{n-2} + \cdots + a_0 2^0$ and $U_x$ as

$$U_x^a = U_x^{a_{n-1}2^{n-1}+a_{n-2}2^{n-2}+\cdots+a_0 2^0}, \tag{8.78}$$

Eq. (8.77) is rewritten to be

$$U_x^a|1\rangle = |x^a \bmod N\rangle = |x^{a_{n-1}2^{n-1}+a_{n-2}2^{n-2}+\cdots+a_0 2^0} \bmod N\rangle. \tag{8.79}$$

Using the modular multiplication (Chapter 12, (A.10)), we can convert Eq. (8.79) into the form

$$\begin{aligned} & x^{a_{n-1}2^{n-1}+a_{n-2}2^{n-2}+\cdots+a_0 2^0} \bmod N \\ & = (x^{a_{n-1}2^{n-1}} \bmod N) \times (x^{a_{n-2}2^{n-2}} \bmod N) \times \cdots \times (x^{a_0 2^0} \bmod N), \end{aligned} \tag{8.80}$$

where the modular operation mod $N$ is applied also to each multiplication on the right-hand side. Thus the algorithm for modular exponentials is given by

**Algorithm for Modular Exponentials $|a\rangle U_x^a|1\rangle = |a\rangle|x^a \bmod N\rangle$**

**Step 1**:

Square $x \bmod N$ to obtain $x^2 \bmod N$.

**Step 2**:

Repeat Step 1 by $(n-2)$ times to obtain

$$x^{2^2} \bmod N, \ldots, x^{2^{n-1}} \bmod N. \tag{8.81}$$

**Step 3**:

For each modular exponential in Eq. (8.81), repeat the congruence operations to obtain

$$\begin{aligned} & (x^{a_{n-1}2^{n-1}} \bmod N)(x^{a_{n-2}2^{n-2}} \bmod N)\cdots(x^{a_0 2^0} \bmod N) \\ & \quad = x^a \bmod N. \end{aligned} \tag{8.82}$$

Let us see the effectiveness of the quantum algorithm for the calculation of modular exponentials. Step 1 consists of the multiplication of two numbers where the amount of computation is $O(n^2)$ at most. In Step 2, this multiplication is repeated by $\sim n$ times with the amount of computation $O(n)$. Therefore the total number of operations in Step 1 and Step 2 is $O(n^3)$ at most. The operation of $x^{2^k} \bmod N$ involves the computation of $O(n^2)$ as shown in Step 1. Since this operation is repeated $\sim n$ times, Step 3 has computation of $O(n^3)$ at most. Hence the total amount of computation with Steps $1 \sim 3$ is $O(n^3)$, namely, the calculation of modular exponentials is an effective algorithm with polynomial complexity.

**Summary of Chapter 8**

(1) The Deutsch-Jozsa algorithm shows the power of the quantum computer.
(2) Shor's algorithm can be used to find the prime factors of a large integer in a polynomial computational time based on congruence equations.
(3) Shor's factorization algorithm makes use of Fourier transformation, phase estimation and order-finding algorithm. There are quantum algorithms to solve these problems in a fast way, i.e., in a polynomial computational time.

## 8.7 Problems

[1] Using the multiplication theorem of congruence equations, prove

$$x^a \bmod N = (x^{a_{n-1}2^{n-1}} \bmod N) \times (x^{a_{n-2}2^{n-2}} \bmod N) \times \cdots \times (x^{a_0 2^0} \bmod N), \qquad (8.83)$$

where mod$N$ is applied also to the products of modular exponentials on the right-hand side.

[2]

**Discrete logarithm problem**

Find an integer $s$ that satisfy

$$a^s \equiv b \bmod N \qquad (8.84)$$

for a prime number $N$ and positive integers $a$ and $b$ less than $N$.

Propose an effective algorithm for the discrete logarithm problem using the following Oracle operator $U$ and the quantum Fourier transformation. The Oracle operator acts as

$$U|x_1\rangle|x_2\rangle|0\rangle = |x_1\rangle|x_2\rangle|f(x_1, x_2)\rangle, \qquad (8.85)$$

where $f$ is a function which gives the value $f(x_1, x_2) = b^{x_1}a^{x_2}\text{mod}N$ to the third register.

Chapter 9

# Quantum Cryptography

From the period of old Roman empire, the cryptography has been used for various purposes including those in military, in diplomacy, and in trade of commerce. Encoding and decoding cryptographic codes have close relation with many types of nation-wide crises and criminal cases in history. Up to the present days, the intelligence agencies of many countries continue to be engaged in developing and breaking new cryptography codes day and night. The RSA cryptography, one of the most sophisticated cryptographies today, is considered to be broken sooner or later when the speed of computers develops remarkably. The cryptography systems in use at present are generally called classical cryptography. Recently, new type of cryptography that is completely different from the classical cryptography is proposed. That is the quantum cryptography. The coding in quantum cryptography is based on the theory of observation in quantum mechanics, and it is considered impossible to tap the codes theoretically due to the irreversible phenomenon called the contraction of wave packets. The implementation of quantum cryptography coding will mean the perfect cryptography that is in principle unbreakable.

**Keywords: secret key cryptography (Caesar cipher,one-time pad cryptography); public key cryptography (RSA cryptography); non-cloning theorem; quantum key distributions (BB84 protocol, B92 protocol, E91 protocol).**

## 9.1 Secret key cryptography

The oldest well-known cryptography is the **Caesar cipher** which is believed to have been invented by Caesar in the period of Roman empire. In the

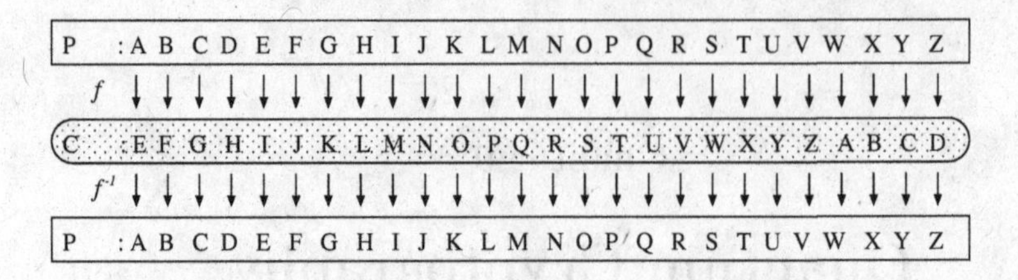


Fig. 9.1 Encoding key$f$ and Decoding key $f^{-1}$ in Caesar cipher.

Caesar cipher, one rewrites each character of the message by shifting a fixed unit in the sequence of alphabets.

Suppose Alice to send a message (plain text) by a cipher text which is encrypted by a key $f$. She sends this cipher text to Bob. Bob, on the other hand, using the decryption key $f^{-1}$, decrypts the cipher text to read the message. Note that Alice must not give the encryption key $f$ to any person other than Bob. Once the encryption key is given to any other person, the decryption key will be made easily, and the coded messages are tapped. This is the weakness of secret key cryptosystems.

For example, suppose that Bob receives the coded message:

$$M\ PSZI\ CSY. \tag{9.1}$$

In order to read this message, he needs to know the key to decrypt it. The encryption key that Alice uses is "4", i.e., each character is shifted by 4 units in the sequence of alphabets. The decryption key to read the message is "−4", i.e., he should shift the characters by 4 units in backward in the alphabets. Thus he can read the message:

$$I\ LOVE\ YOU. \tag{9.2}$$

In the Caesar cipher, the secrecy of the key is one of the most important subjects because messages will be tapped as soon as the secret key is known.

By mapping the 26 alphabets $A$, $B$, $C$, ..., $Z$ on the sequence of 26 numbers from 0 to 25, Caesar cipher is expressed by the following function $f$ with respect to $x \in \{0, 1, \ldots, 25\}$:

$$f(x) = \begin{cases} x + \alpha & \text{if } x < 26 - \alpha, \\ x + \alpha - 26 & \text{if } x \geq 26 - \alpha, \end{cases} \tag{9.3}$$

where $\alpha$ is the encryption key and is a number between 1 and 25. In (9.1), $\alpha = 4$. Equation (9.3) can be also expressed by the notation of congruence equations as

$$f(x) \equiv x + \alpha \pmod{26} \tag{9.4}$$

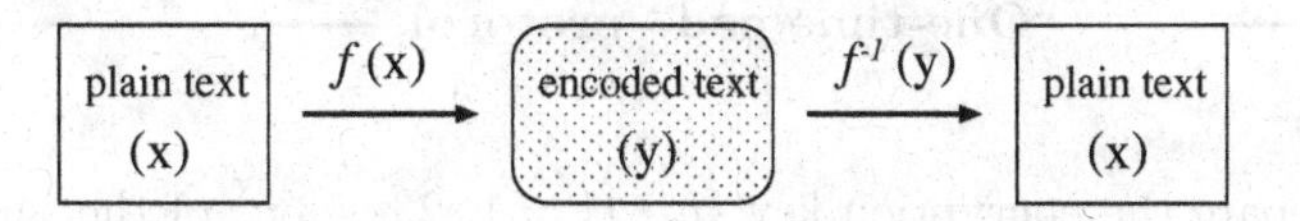


Fig. 9.2 Encryption and decryption in Caesar cipher.

with modulus 26. The function $f^{-1}$ is a single-valued function. The decryption key to recover the original message from the encrypted one is given by

$$f^{-1}(y) \equiv y - \alpha \pmod{26}. \tag{9.5}$$

## 9.2 "One-time pad" cryptography

**"One-time pad" cryptography** was proposed around the end of the World War I. The name "one-time pad" came from "note pad to use only once", and the system uses random patterns, such as random numbers. The key is used only once, and is never used again. Let us suppose that Alice send a plain text of length $n$,

$$(a_1, a_2, \ldots, a_n) \tag{9.6}$$

to Bob by encrypting it by one-time pad system. In advance, they prepare a random pattern of the same length,

$$(b_1, b_2, \ldots, b_n) \tag{9.7}$$

as the secret key, which Alice and Bob share. Alice makes the codes $c_i$ by the congruence equation:

$$c_i \equiv a_i + b_i \pmod{N} \tag{9.8}$$

modulo a certain integer $N$. Bob, having the same secret key $b_i$, can decrypt them by the congruence equation:

$$a_i \equiv c_i - b_i \pmod{N} \tag{9.9}$$

to obtain the original plain text. In the Caesar cipher of the example, $N = 26$ and the key for encrypting $b_i$ is a constant $\alpha$. If we express the pattern in $N = 2$ bits, and Eqs. (9.8) and (9.9) reduce to exclusive OR:

$$c_i \equiv a_i \oplus b_i \tag{9.10}$$

$$a_i \equiv c_i \oplus b_i. \tag{9.11}$$

**"One-time pad" protocol**

**Step 1**

(Alice) : Prepare the encryption key $\{b_i\}$ $(i = 1, 2, \cdots, n)$ of the same length as the plain text$\{a_i\}$ $(i = 1, 2, \cdots, n)$. Share it with Bob as the secret key.

**Step 2**

(Alice) : Encrypt the plain text using the key and send the cipher to Bob.

$$c_i \equiv a_i + b_i \pmod{N} \tag{9.12}$$

**Step 3**

(Bob) : Decode the cipher using the key.

$$a_i \equiv c_i - b_i \pmod{N} \tag{9.13}$$

"One-time pad" cryptosystem has the following characteristics:

(1) The encryption key consists of random patterns, or random numbers.
(2) The key is used only once, i.e., the key is changed every time.

From these characteristics, it overcomes the weakness of the old secret key system, and as a result it is an unbreakable cryptosystem. "One-time pad" cryptography, which looked safe theoretically, acquired little use. The major shortcomings in practice are

(1) A large number of random patterns must be generated as secret keys.
(2) A large number of secret keys must be distributed without being wire-tapped.

About the first shortcoming, it is quite hard, even with the present technology, to generate tremendous random patterns or numbers. For the second, it will be also extremely difficult because so many different keys must be distributed. Recently, however, it has turned out that these two shortcomings can be overcome by the quantum cryptosystem. As a result, "one-time pad" cryptosystem is being revived.

## 9.3 Public key cryptography

Like the Caesar cipher described in Section 9.1, the widely used cryptosystems so far belong to **Secret Key Cryptography**, in which the encryption key and the decryption key are secret known only to the sender (Alice) and the receiver (Bob), and the information about them must be strictly controlled. On the contrary, once the information about these keys leaks, the ciphers are broken quite easily. All the secret key cryptographies like Caesar cipher have been broken by experts. There is more advanced cryptography, which called the public key cryptosystem.

In 1976, W. Diffie and M. Hellman invented a cryptography theory that is completely different from conventional ones. That is the cryptosystem called **public key cryptography**. In public key cryptosystem, the encryption key is made open to the public, therefore anyone can access to it. The decryption key, on the other hand, can not be obtained easily even if the encryption key is known. Among cryptosystems based on the public key, one of the most widely used systems is called **RSA cryptography**.

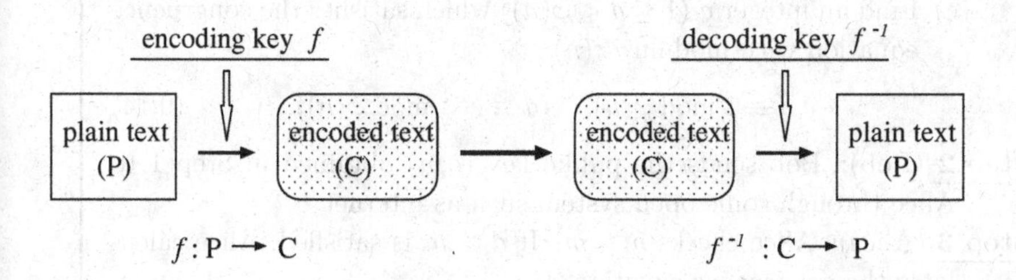


Fig. 9.3 Encryption and decryption.

The idea of public key cryptography is similar to that of one-way functions. With a one-way function, it is easy to obtain $f(x)$ for a given $x$ while it is impossible to obtain $x$ from a given $f(x)$. In public key cryptography, it is "almost impossible" to know the decryption key even if one has the encryption key, as explained below. The words "almost impossible" means that the amount of computation is so big that it would require the computation time of the order of the age of the universe (billions of years) with the fastest computers.

The name RSA was taken after the inventors R. Rivest, A. Shamir and L. Adleman. RSA cryptography is based on the fact that the products of two large numbers are obtained easily, but there is no effective algorithm for the factorization of big numbers. In the following, we describe how to make ciphers by RSA cryptography. Note that the RSA cryptography was invented entirely based upon deep knowledge of number theory.

**RSA cryptography protocol**

Let $m$ the message (in practice we take a large integer with more than 300 digits). We are going to make a public key and a secret key for the encryption of this message in RSA cryptography.

**Step 1** (Bob): To make a public encryption key $(e, n)$ and a secret decryption key $(d, n)$ from two natural numbers:

a) Take a pair of big prime numbers $p$ and $q$ so that the product $pq$ is larger than $m$. Calculate the product $n = pq$ and the Euler function[1] of it $\varphi(n) = (p-1)(q-1)$. The inequality $n > m$ must be satisfied, though this condition can be checked later by Alice when Bob sends her the public key $(e, n)$.

b) Take an integer $e$ which is less than $n$ and satisfies the equation $\gcd(\varphi(n), e) = 1$.

c) Find an integer $d$ ($1 \leq d < \varphi(n)$) which satisfies the congruence equation with modulus $\varphi(n)$

$$e\,d \equiv 1 \bmod \varphi(n) \qquad (d \equiv e^{-1} \bmod \varphi(n)). \tag{9.14}$$

**Step 2** (Bob): Bob sends the public key $(e, n)$ obtained in Step 1 to Alice through some open system such as internet.

**Step 3** (Alice): Alice checks $n > m$. If $n > m$ is satisfied, Alice calculates the congruence equation

$$m^e \equiv y \bmod n, \tag{9.15}$$

from her message $m$ and the public key $(e, n)$, and sends the result $y$ as the cipher to Bob.

**Step 4** (Bob): Bob calculates

$$y^d \bmod n. \tag{9.16}$$

using the cipher $y$ from Alice and the secret key $(d, n)$. The result of Eq. (9.16) provides Alice's message $m$, as is proved with the help of Euler's theorem (See Appendix).

A characteristic of RSA cryptography lies in the point that it is hard to obtain the decryption key $d$ from the the encryption key $e$. In other words, to get the decryption key $d$, one needs to have the Euler function

[1] For a given natural number $n$, the Euler function $\varphi(n)$ gives the natural numbers that are less than and prime to $n$. In particular, if $n$ is a prime number, $\varphi(n) = n - 1$.

$\varphi(n)$, i.e., one needs to factorize the modulus $n$ in two prime numbers to obtain $\varphi(n)$. The factorization is difficult as the modulus $n$ gets bigger. In practice, such large numbers as $n \gtrsim 10^{300}$ ($\log_2 n \gtrsim 1000$) are adopted to guarantee sufficient security. The factorization of a number of 300 digits would need $\sqrt{N}$ times of operation, i.e., $10^{150}$ operations would be needed. It would take time of the order of the age of the universe (billions of years) even with the fastest computers available now.

---

**Example 9.1** Prove the congruence equation $y^d \equiv m(\text{mod } n)$ of (9.16).

**Solution** From (9.14), we can write $e\,d$ as

$$e\,d = 1 + k\varphi(n) \tag{9.17}$$

using some integer $k$. The expression $y^d(\text{mod } n)$ can be written as

$$y^d \equiv m^{d\,e} \equiv m \cdot m^{k\varphi(n)}(\text{mod } n). \tag{9.18}$$

using Eqs. (9.15) and (9.17). Now, Euler's theorem $m^{\varphi(n)} \equiv 1(\text{mod } n)$ gives $m^{k\varphi(n)} \equiv 1(\text{mod } n)$ and

$$y^d \equiv m(\text{mod } n).[-6pt] \tag{9.19}$$

---

---

**Example 9.2** Encrypt $m = 77$ by RSA cryptography, and decrypt the cipher obtained.

**Solution**

**Step 1** :

a) (Bob): Take sufficiently large two primes $p$ and $q$, and calculate their product $n$. Let us choose $p = 5$ and $q = 17$, which gives We have

$$n = 5 \times 17 = 85 > m. \tag{9.20}$$

Thus the condition $n > m$ is satisfied. The Euler function is

$$\varphi(n) = \varphi(85) = \varphi(5)\varphi(17) = 4 \times 16 = 64 = 2^6. \tag{9.21}$$

b) Choose $e = 3$ which satisfies $\gcd(\varphi(n), e) = 1$.

c) Find $d$ satisfying both $1 \leq d \leq \varphi(n)$ and

$$e\,d = 3d = 1 \text{ mod } (\varphi(n)). \tag{9.22}$$

Since $\varphi(n) = \varphi(85) = 64$, Eq. (9.22) can be expressed as

$$3d - 1 = 64k \quad (k \text{ is an integer}). \tag{9.23}$$

By writing this equation as

$$3d - 64k = 1, \tag{9.24}$$

we have a linear indefinite equation, one of the Diophantus equations. The general solution of the Diophantus equations obtained as follows. The relation using Euclid's algorithm (See Appendix),

$$-3 \times 21 + 64 = 1, \tag{9.25}$$

gives a set of solutions $d_0 = -21$ and $k_0 = -1$ of Eq. (9.24). Then the general solution can be given as

$$\begin{matrix} d = -21 + 64t \\ k = -1 + 3t. \end{matrix} \quad (t \text{ is an arbitrary integer}). \tag{9.26}$$

If we fix $t = 1$, then $d = 43$. In this way, Bob has generated the public encryption key $(e, n) = (3, 85)$ and the secret decryption key $(d, n) = (43, 85)$.

**Step 2** :

Bob sends the public key: $(e, n) = (3, 85)$ to Alice.

**Step 3** :

Alice uses the public key to encrypt the message $m$ (plain text). The congruence equation

$$77^3 \equiv 83 \bmod 85, \tag{9.27}$$

provides the cipher $y = 83$. Alice sends this cipher $y = 83$ to Bob.

**Step 4** :

To decrypt the cipher $y = 83$ by the secret key $(d, n) = (43, 85)$, Bob calculates the congruence equation

$$y^{43} = 83^{43} \tag{9.28}$$

modulo 85.

The congruence equation can be solved by calculating $83^n \bmod 85$ with increasing the value of $n$ ($n = 1, 2, 3, \ldots$) until one finds the order $r$. Once one finds the order $r$,

$$83^r \equiv 1 \bmod 85, \tag{9.29}$$

one can make use of the theorem for a positive integer $\alpha$,

$$83^{\alpha r} \equiv 1 \bmod 85, \tag{9.30}$$

to decrypt easily the cipher even if $d$ is large. From

$$\begin{aligned} 83^2 &\equiv 4 \bmod 85 \\ 83^4 &\equiv 16 \bmod 85 \\ 83^8 &\equiv 256 \equiv 1 \bmod 85 \end{aligned} \tag{9.31}$$

we find the order $r$=8. Choosing $\alpha = 5$, we can calculate

$$\begin{aligned} 83^{5\times 8} &\equiv 83^{40} \equiv 1 \bmod 85 \\ 83^{43} &\equiv 83^{40}83^3 \equiv 83^3 \equiv 77 \bmod 85, \end{aligned} \tag{9.32}$$

and we finish the decryption with the result: $m = 77$

---

The point of RSA cryptography is that decryption is impossible without the knowledge of the Euler function $\varphi(n)$, even if both the public key $(e, n)$ and the cipher $y$ are known. To obtain the Euler function, one needs to factorize $n$ into two prime numbers $p$ and $q$. In Example 9.2, the factorization was rather easy because $n = 85$ was quite small. How about the case with $p$ and $q$ with 100 digits or more, and prime to each other, i.e., $n$ has more than 200 digits? No powerful algorithm for factorization has been discovered yet on the conventional computers of the present days. Since the algorithm can not be finished within the polynomial time. it would take astronomical length of time, namely, the time comparable to the lifetime of the universe to solve the problem for such large integers, i.e., it is practically impossible to solve. However, as was described in Chapter 8, if the quantum computer is available, Shor's algorithm can perform factorization in polynomial times. Then the RSA cryptography may not be a safe scheme. While Shor's quantum algorithm will make it easy to break RSA ciphers, other types of invincible cryptography are proposed based on the principles of quantum mechanics. That is the quantum cryptography.

## 9.4 Quantum key distribution

There are two major schemes to distribute the key in the conventional cryptography. One is to send the keys through public channels with sufficient reliability. For example, one can keep secret by distributing the keys only to the proper persons with the help of secret agency. The other scheme is the public key distribution. Neither of these two schemes are, however, absolutely safe. For the first method, it often happens that the messengers betray. About the second, its security is broken if an efficient factorization algorithm is implemented. In particular, if Shor's factorization algorithm is put in practice by a quantum computer, RSA cryptography can not stay safe any more. On the other hand, it turns out that the uncertainty principle of quantum mechanics will detect the existence of tappers in the process of key distribution, and proposes a new type of cryptography. These quantum cryptography theories such as BB84 protocol, B92 protocol and E91 protocol will be described in the following.

### 9.4.1 *Non-cloning theorem*

In distributing the keys for cryptography in quantum-mechanical way, the guarantee of secrecy is based on the observation theory in quantum mechanics, namely, the quantum bits can never be tapped. This property is often called the non-cloning theorem. Sheep can be cloned, but quantum bits can never be cloned.

---

**Example 9.3** Show that a qubit $|\psi\rangle$ can not be cloned

**Solution** We think about how to copy a quantum bit $|\psi\rangle$ to a target bit $|s\rangle$. The initial state is given by

$$|\psi\rangle \otimes |s\rangle. \tag{9.33}$$

Suppose that we apply a unitary operator $U$ to this state to copy $|\psi\rangle$ to the target bit $|s\rangle$:

$$U(|\psi\rangle \otimes |s\rangle) = |\psi\rangle \otimes |\psi\rangle. \tag{9.34}$$

If we applying the copying operator $U$ to another quantum bit $|\phi\rangle$, we have

$$U(|\phi\rangle \otimes |s\rangle) = |\phi\rangle \otimes |\phi\rangle. \tag{9.35}$$

Now let us make a scalar product of Eq. (9.34) and Eq. (9.35). Because $U$ is unitary, the product of the left-hand side becomes

$$\langle s| \otimes \langle \phi|U^{-1}U|\psi\rangle \otimes |s\rangle = \langle \phi|\psi\rangle. \tag{9.36}$$

The right-hand side, on the other hand, becomes

$$\langle \phi| \otimes \langle \phi|\psi\rangle \otimes |\psi\rangle = \langle \phi|\psi\rangle^2. \tag{9.37}$$

From Eqs. (9.36) and (9.37), we have

$$\langle \phi|\psi\rangle = \langle \phi|\psi\rangle^2. \tag{9.38}$$

Therefore the cloning is possible only if the scalar product of $\phi$ and $\psi$ satisfies

$$\langle \phi | \psi \rangle = 1 \quad \text{or} \quad 0. \tag{9.39}$$

This result shows that the cloning is possible only if $|\phi\rangle = |\psi\rangle$, or $|\phi\rangle$ is orthogonal to $|\psi\rangle$. In general, the qubits are linear combinations of binary bits, $|\psi\rangle = a|0\rangle + b|1\rangle$ and $|\phi\rangle = c|0\rangle + d|1\rangle$, so that they are not orthogonal to each other,

$$0 < |\langle \phi | \psi \rangle| = |c^* a + d^* b| < 1. \tag{9.40}$$

Hence it is impossible to make a cloning operator for qubits.

### 9.4.2 *BB84 protocol*

The Quantum key distribution protocol making use of polarized photons was proposed by C. H. Bennett and G. Brassard for the first time in 1984. The protocol for cryptography is called **BB84 protocol** after the initials of the two inventors and the year.[2] The BB84 protocol is the system which combines the theory of measurement in quantum mechanics and the idea of the cryptography based on one-time pad, making the cryptography that can never be tapped.

In the BB84 protocol, one prepares four kinds of polarized photons by laser: rectilinearly polarized photons, vertically and horizontally, $|\updownarrow\rangle$ and $|\leftrightarrow\rangle$; and diagonally polarized photons, at angle $+45°$ and $-45°$, $|\nearrow\rangle$ and $|\nwarrow\rangle$. The binary bits 0 and 1 are expressed by two kinds of polarized photon, as shown in Table 9.1.

For example, a number 22 is expressed as 10110 in the binary system, and the representation by the rectilinear polarizations, is

$$|\leftrightarrow\rangle |\updownarrow\rangle |\leftrightarrow\rangle |\leftrightarrow\rangle |\updownarrow\rangle, \tag{9.41}$$

Table 9.1 Representations of binary values 0 and 1 by two kinds of polarized photons.

| bit value | $\oplus$ | $\otimes$ |
|---|---|---|
| 0 | $\lvert\updownarrow\rangle$ | $\lvert\nearrow\rangle$ |
| 1 | $\lvert\leftrightarrow\rangle$ | $\lvert\nwarrow\rangle$ |

[2]C. H. Bennett and G. Brassard, Proc. of IEEE Int. Conference on Computers, Systems and Signal Processing, p. 175 (IEEE, New York, 1984).

while the representation by the diagonal polarizations is

$$|\searrow\rangle|\swarrow\rangle|\searrow\rangle|\searrow\rangle|\swarrow\rangle. \tag{9.42}$$

Let us assume that Alice and Bob communicate by the rectilinearly polarized photons only. Suppose that the tapper uses the same polarization filter $\oplus$ as Alice, and sends the polarized photons to Bob. Then the tapper succeeds in tapping the message without being noticed by either Alice nor Bob. Therefore, the communication by only one set of polarized photons has no guarantee for the security against tappers. In fact, it is possible to clone two mutually orthogonal states.

The characteristic of the BB84 protocol lies in using two sets of polarization, rectilinear and diagonal. To be more specific, Alice uses two kinds of filters $\oplus$ and $\otimes$ in a random order to send bits, while Bob receives the photons using two detectors $\oplus$ and $\otimes$ in a random order.

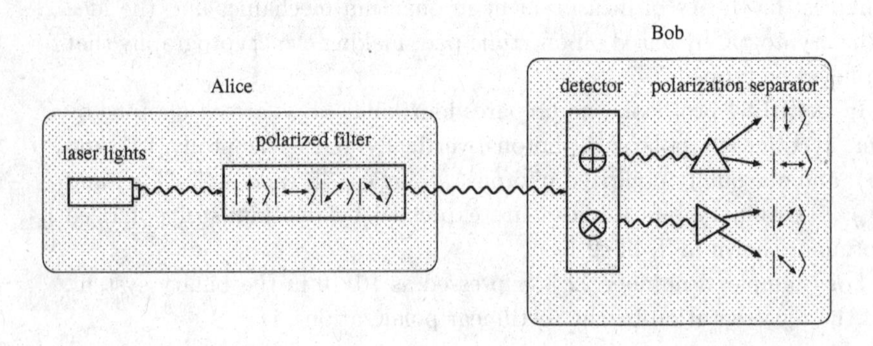


Fig. 9.4 Communication by the BB84 protocol. Alice operates the laser device and the polarization filters choosing one of the directions vertical/horizontal or diagonal in a random order to send the bits 0 or 1. Bob observes the direction of photons choosing the detector and the polarization separator for rectilinear $\oplus$ or diagonal $\otimes$ directions.

Let us follow the procedure to send an $n$-bit quantum key in the BB84 protocol.

**BB84 protocol**

**Step 1** :

Alice chooses one of two polarization filters $\oplus$ or $\otimes$ randomly to send Bob the data of the random sequence of $4n$ bits.

**Step 2** :

Bob chooses one of the two polarization detectors $\oplus$ or $\otimes$ randomly to observe the directions of the polarized photons sent by Alice. Through some classical communication channel such as telephone or internet, Alice let Bob know only the order of polarization filters that she has chosen herself. Note that she never told him anything about the random data of $4n$ bits.

**Step 3** :

The both party select only the results of the measurement in the cases where the type of the detector agrees with the type of the polarization filter. Since the probability that Bob has the same kind of the polarization for the detectors as that for Alice's photon filers is $1/2$, $2n$ bits are expected to be in common among $4n$ bits. Out of those $2n$-bit data, they will compare $n$ bits. If the $n$ bits with Bob agree with corresponding $n$ bits with Alice perfectly, they adopt the remaining $n$-bit data as the one-time pad. If the agreement is not perfect, the tapper is suspected. In this case, they give up all the present data in hand and repeat the procedures from Step 1 as the next session.

**Stop 4** :

Alice encrypts her message using the one-time pad generated in Step 3 and send the cipher to Bob.

**Stop 5** :

Bob decrypts the cipher from Alice using the common one-time pad .

In these two kinds of polarization, the rectilinear photons are expressed as

$$|0\rangle_{\oplus} = |\updownarrow\rangle \tag{9.43}$$

$$[9pt]|1\rangle_{\oplus} = |\leftrightarrow\rangle, \tag{9.44}$$

while the diagonal states can be expressed as superpositions of the rectilinear basis,

$$|0\rangle_{\otimes} = \frac{1}{\sqrt{2}}\{|\updownarrow\rangle + |\leftrightarrow\rangle\}, \tag{9.45}$$

$$|1\rangle_{\otimes} = \frac{1}{\sqrt{2}}\{|\updownarrow\rangle - |\leftrightarrow\rangle\}. \tag{9.46}$$

Equations (9.45) and (9.46) show that when one measures the diagonal polarization states with the rectilinear detector, one finds the state $|\updownarrow\rangle$ or the state $|\leftrightarrow\rangle$ with the probability of 1/2. Therefore, if Alice sends the information $|0\rangle_{\otimes}$, Bob will find $|0\rangle_{\oplus}$ or $|1\rangle_{\oplus}$ with the probability 1/2 each if he uses the rectilinear detector, $\oplus$. As a result, Bob will receive the information from Alice incorrectly with probability of 50%.

Table 9.2 BB84 protocol. Alice chooses bit values and the polarization filter randomly. Bob chooses the polarization detector randomly, too. Alice and Bob exchange the information about their polarizations which they have chosen through any public communication channel like telephone or Internet. They adopt the pattern of the bits with common polarizations as the one-time pad.

| | | 1 | 2 | 3 | 4 | 5 | 6 | 7 | 8 | 9 | 10 | 11 | 12 |
|---|---|---|---|---|---|---|---|---|---|---|---|---|---|
| Alice | value | 0 | 1 | 0 | 1 | 1 | 0 | 0 | 1 | 1 | 1 | 0 | 1 |
| | filter | $\otimes$ | $\oplus$ | $\oplus$ | $\otimes$ | $\otimes$ | $\oplus$ | $\otimes$ | $\oplus$ | $\oplus$ | $\otimes$ | $\otimes$ | $\otimes$ |
| | polarization | $\swarrow\!\!\!\nearrow$ | $\leftrightarrow$ | $\updownarrow$ | $\nwarrow\!\!\!\searrow$ | $\nwarrow\!\!\!\searrow$ | $\updownarrow$ | $\swarrow\!\!\!\nearrow$ | $\leftrightarrow$ | $\leftrightarrow$ | $\nwarrow\!\!\!\searrow$ | $\swarrow\!\!\!\nearrow$ | $\nwarrow\!\!\!\searrow$ |
| Bob | detector | $\otimes$ | $\oplus$ | $\otimes$ | $\oplus$ | $\oplus$ | $\oplus$ | $\otimes$ | $\otimes$ | $\oplus$ | $\oplus$ | $\oplus$ | $\otimes$ |
| | polarization | $\swarrow\!\!\!\nearrow$ | $\leftrightarrow$ | $\nwarrow\!\!\!\searrow$ | $\updownarrow$ | $\leftrightarrow$ | $\updownarrow$ | $\swarrow\!\!\!\nearrow$ | $\updownarrow$ | $\leftrightarrow$ | $\leftrightarrow$ | $\updownarrow$ | $\nwarrow\!\!\!\searrow$ |
| | value | 0 | 1 | 1 | 0 | 1 | 0 | 0 | 0 | 1 | 1 | 0 | 1 |
| agreement of filter | | T | T | F | F | F | T | T | F | T | F | F | T |
| one-time pad | | 0 | 1 | | | | 0 | 0 | | 1 | | | 1 |

As shown in Table 9.2, Alice sends the signal of the random sequence of bits 010110011101 with random polarization filters and Bob also choose the polarization detectors randomly. As a result, he observes the signal of 011010001101. As seen from Table 9.2, even in the cases where Bob uses the detector with different polarization from Alice, he detects correct values of bits with probability 50%. In Step 3, Alice and Bob exchange information through Internet or telephone. The information is restricted to the polarization that they have chosen. Note that the secrecy of the bits sent will be kept perfectly secured, in this way, even if the Internet is tapped. In Step 4, from only those cases where the types of polarizations with Alice and Bob agrees, namely, from the 1, 2, 6, 7, 9 and 12th bits, they can generate one-time pad 010011. Using this bit sequence of one-time pad, Alice makes the cipher and send it to Bob. By using random sequences in this way, one can make ciphers which is impossible to be tapped. In conventional cryptography with one-time pad, the most troublesome problem has been how to send the random sequence securely without being tapped. By using the BB84 protocol, we can make the one-time pad, which is impossible to be tapped, under the principles of quantum mechanics.

Now, assuming a tapper, Eve, we will show how we can discover her presence by the protocol. If Eve tries to steal the bit sequence that Alice sends, the only way that she can take will be to choose the detectors in a random way like Bob did. Then, in the example of Table 9.3, she observes the wrong bit sequence 1, 4, 12 as a result of her random choice of polarization types According the the theory of measurement in quantum mechanics, this kind of errors are unavoidable as long as the observer has no knowledge about the types of polarization filters that Alice chooses for each bit. We assume that Eve sends Bob the same polarization as she observes. When Bob measures the state of polarization, he will obtain, for example, the result in Table 9.3. Due to Eve's measurement of Alice's data, the quantum states has been disturbed, and consequently the one-time pad 010011 adopted as the encryption key was distorted into 110001. In this way, even if the receiver has the same polarization as the sender, the disturbance by the tapper causes the difference between the transmitted bits and the received bits, and finally the existence of the tapper will be known.

Table 9.3 The bit sequence that Bob observes in the data disturbed by the tapper "Eve".

| | | 1 | 2 | 3 | 4 | 5 | 6 | 7 | 8 | 9 | 10 | 11 | 12 |
|---|---|---|---|---|---|---|---|---|---|---|---|---|---|
| Alice | value | 0 | 1 | 0 | 1 | 1 | 0 | 0 | 1 | 1 | 1 | 0 | 1 |
| | filter | ⊗ | ⊕ | ⊕ | ⊗ | ⊗ | ⊕ | ⊗ | ⊕ | ⊕ | ⊗ | ⊗ | ⊗ |
| | polarization | ⤢ | ↔ | ↕ | ⤡ | ⤡ | ↕ | ⤢ | ↔ | ↔ | ⤡ | ⤢ | ⤡ |
| Eve | detector | ⊕ | ⊗ | ⊗ | ⊕ | ⊕ | ⊕ | ⊗ | ⊕ | ⊗ | ⊗ | ⊗ | ⊕ |
| | observation | ↔ | ⤡ | ⤢ | ↕ | ↔ | ↕ | ⤢ | ↔ | ⤡ | ⤡ | ⤢ | ↕ |
| | value | 1 | 1 | 0 | 0 | 1 | 0 | 0 | 1 | 1 | 1 | 0 | 0 |
| Bob | detector | ⊗ | ⊕ | ⊗ | ⊕ | ⊕ | ⊕ | ⊗ | ⊗ | ⊕ | ⊕ | ⊕ | ⊗ |
| | observation | ⤡ | ↔ | ⤢ | ↕ | ↔ | ↕ | ⤢ | ⤡ | ↕ | ↕ | ↕ | ⤡ |
| | value | 1 | 1 | 0 | 0 | 1 | 0 | 0 | 1 | 0 | 0 | 0 | 1 |
| one-time pad | | 1 | 1 | | | | 0 | 0 | | 0 | | | 1 |

Let use estimate the probability of Alice and Bob finding the tapper by comparing the $n$ bits in one-time pad. When Alice and Bob have the same type of polarization, Bob will observe an incorrect value if Eve uses the other type of the detector and furthermore Bob observes the value opposite to one Alice has sent. The probability of Eve using the other detector is $\frac{1}{2}$, and that of Bob observing the opposite value is also $\frac{1}{2}$. Therefore the probability of finding the tapping by comparing a single bit is $\frac{1}{2} \times \frac{1}{2} = \frac{1}{4}$. Thus the probability of finding tapping by checking $n$ times is

$$P(n) = 1 - \left(\frac{3}{4}\right)^n . \tag{9.47}$$

If they check with 10 bits, then $P(10) = 0.939$, and with 20 bits, $P(20) = 0.996$. In this way, we know that the presence of a tapper will be revealed with very large probability.

---

**Example 9.4** Show that the probability of revealing the existence of a tapper by comparing $n$ bits by the BB84 protocol is given by Eq. (9.47),

$$P(n) = 1 - \left(\frac{3}{4}\right)^n.$$

**Solution** For $n = 1$, the probability $P(1) = 1 - \frac{3}{4} = \frac{1}{4}$ has already derived from the probability of the disagreement of the randomly chosen polarization detectors and the probability with the measurement of quantum states. For $n = 2$, note that the probability of the two bits having different values is $\frac{1}{4} \times \frac{1}{4}$, while the probability of a bit having the same value and the other having different values is $2 \times \frac{3}{4} \times \frac{1}{4}$. Therefore we have

$$P(2) = \frac{1}{4} \times \frac{1}{4} + 2 \times \frac{3}{4} \times \frac{1}{4} = \frac{7}{16} = 1 - \left(\frac{3}{4}\right)^2. \tag{9.48}$$

For the case with $n$ bits, the probability of $n$ bits having different values is $\left(\frac{1}{4}\right)^n$, that of $(n-1)$ bits having different values is $\frac{3}{4}\left(\frac{1}{4}\right)^{n-1} n, \cdots$, that of 1 bit having different values is $\left(\frac{3}{4}\right)^{n-1} \frac{1}{4} n$. By summing them, we have

$$\begin{aligned} P(n) &= \left(\frac{1}{4}\right)^n + \frac{3}{4}\left(\frac{1}{4}\right)^{n-1} n + \cdot + \left(\frac{3}{4}\right)^{n-1} \frac{1}{4} n \\ &= \sum_{k=0}^{n-1} \binom{n}{k} \left(\frac{1}{4}\right)^{n-k} \left(\frac{3}{4}\right)^k = \sum_{k=0}^{n} \binom{n}{k} \left(\frac{1}{4}\right)^{n-k} \left(\frac{3}{4}\right)^k - \left(\frac{3}{4}\right)^n \\ &= \left(\frac{1}{4} + \frac{3}{4}\right)^n - \left(\frac{3}{4}\right)^n = 1 - \left(\frac{3}{4}\right)^n. \end{aligned} \tag{9.49}$$

In an alternative way, one can obtain the solution considering the probability not to be able to find a tapper. Notice that each bit has the same value between the sender and the receiver with the probability of $3/4$ regardless of a tapper. For this case, the tapper can not be found with the probability of $3/4$. For $n-$bits, the probability unaware of the tapper becomes $(3/4)^n$. Contrarily, the presence tapper can be recognized with the probability of $1 - (3/4)^n$ which is identical to Eq. (9.49).

---

### 9.4.3 *B92 protocol*

In the BB84 protocol, one needs to prepare four kinds of polarized states: two rectilinear polarized states, $|\updownarrow\rangle, |\leftrightarrow\rangle$, and two diagonally polarized states $|\searrow\rangle, |\nearrow\rangle$, together with two types of detectors. The B92 protocol, on the other hand, distributes the secret key with two non-orthogonal states. This method is called B92 because it was proposed by C. H. Bennett in 1992.[3] In the B92 protocol, two non-orthogonal states are assigned to 0 and 1 in the binary system, namely, the vertically polarized state $|\updownarrow\rangle$ represents 0 while the states polarized at $-45°$, $|\searrow\rangle$ represents 1.

Table 9.4 States of polarization and detector to project bit values in B92 protocol.

| Alice | | Bob |
|---|---|---|
| bit value | state of polarization | detector |
| 0 | $\updownarrow$ | $P_0 = 1 - \vert\searrow\rangle\langle\searrow\vert$ |
| 1 | $\searrow$ | $P_1 = 1 - \vert\updownarrow\rangle\langle\updownarrow\vert$ |

From Eqs. (9.43) and (9.46), the scalar product of these two states is calculated as

$$\langle\searrow|\updownarrow\rangle = \frac{1}{\sqrt{2}}. \tag{9.50}$$

The receiver Bob has two kinds of detectors $P_0$ and $P_1$. He obtains the value 0 if detector $P_0$ does not observe the state $|\searrow\rangle$, while he obtains the value 1 if detector $P_1$ does not observe the state $|\updownarrow\rangle$. In other words, if Alice transmit the bit value $|0\rangle = |\updownarrow\rangle$, the detector $P_1$ gives

$$\langle\searrow|P_1|\updownarrow\rangle = \langle\searrow|(|\updownarrow\rangle - |\updownarrow\rangle) = 0. \tag{9.51}$$

On the other hand, because

$$\langle\updownarrow|P_0|\updownarrow\rangle = 1 - |\langle\searrow|\updownarrow\rangle|^2 = 1 - \frac{1}{2} = \frac{1}{2}, \tag{9.52}$$

he will observe the bit value 0 with the probability of 1/2. Similarly, we have

$$\langle\updownarrow|P_0|\searrow\rangle = 0 \tag{9.53}$$

$$\langle\searrow|P_1|\searrow\rangle = \frac{1}{2} \tag{9.54}$$

for the diagonally polarized state $|\searrow\rangle$ corresponding the value 1. Thus, the detector $P_1$ gives value 1 with probability of 1/2. The characteristic of the B92 protocol is that it can deliver the correct bit values without error with probability of 50% using two kinds of photons.

[3] C. H. Bennett, Phys. Rev. Lett. **68**, 3121 (1992).

Table 9.5 Transmission and reception of bit values by B92 protocol. Alice makes use of two non-orthogonal polarized states. Bob arranges bit-projection detectors $P_0$ and $P_1$ randomly, and he adopts as the one-time pad the values of bits of received signals.

| | | 1 | 2 | 3 | 4 | 5 | 6 | 7 | 8 | 9 | 10 | 11 | 12 |
|---|---|---|---|---|---|---|---|---|---|---|---|---|---|
| Alice | value | 0 | 1 | 0 | 1 | 1 | 0 | 0 | 1 | 1 | 1 | 0 | 1 |
| | pol. | $\updownarrow$ | $\searrow$ | $\updownarrow$ | $\searrow$ | $\searrow$ | $\updownarrow$ | $\updownarrow$ | $\searrow$ | $\searrow$ | $\searrow$ | $\updownarrow$ | $\searrow$ |
| Bob | detector | $P_0$ | $P_1$ | $P_1$ | $P_0$ | $P_1$ | $P_0$ | $P_1$ | $P_1$ | $P_0$ | $P_1$ | $P_0$ | $P_0$ |
| | signal | y | n | n | n | n | y | n | y | n | n | y | n |
| one-time pad | | 0 | | | | | 0 | | 1 | | | 0 | |

Table 9.5 shows an illustrative example of the B92 protocol. Alice transmits random bit values to Bob using two kinds of non-orthogonal polarized states. Bob arranges two detectors $P_0$ and $P_1$ to receive the signals. Only those bit values 0010, for which the signals are received, constitutes the secret key for the one-time pad. Bob let Alice know which signals he has received, telling her nothing about the types of detectors. In the example, the 1st, 6th, 8th and 11th signals have been received. He delivers these numbers in some classical way. Even these numbers are tapped by Eve, the secret key of the one-time pad is distributed with 100% security because Eve does not know what kinds of detectors have been used. As clearly seen in Table 9.5, the detector never receives the signal when the types of detector does not fit the transmitted bit values. Only if the two types agree, the detector receives the signal with probability of 1/2. In the case of Table 9.5, the 2nd, 5th and 10th signals are not received in spite of the fact that the bit values are the same as the detectors. If Bob chooses the detectors randomly, the probability of Alice's signals being received is

$$\frac{1}{2}(1 - \langle \updownarrow | \searrow \rangle^2) = \frac{1}{4}. \tag{9.55}$$

Despite this fact, the important point with the B92 protocol is that the probability of incorrect signals being received is zero. If the tapper "Eve" tries to tap the bits that Alice transmits, Eve can not copy the bit from Alice because the non-cloning theorem (see Example 9.3) for the two non-orthogonally polarized states. Therefore, when Eve traps the bits and transmits the observed state again by herself, these procedures leads to the reveal of the tapper. In general, for the communication by BB84 and B92, two kinds of bit patterns must be prepared, i.e., the bits to send the original text and the bits to detect tappers.

### 9.4.4 *E91 protocol*

The BB84 protocol and the B92 protocol are the key distribution methods for which the security is guaranteed from the non-cloning principle of quantum states, namely, the cloning is not possible for the two no-orthogonal quantum states. In these two protocols, the uncertainty principle of quantum mechanics plays an essential role. A. K. Ekert in 1991[4] proposed a key distribution protocol with the "entangled" quantum states so called EPR pairs. Let us now discuss a method to distribute keys by making use of entangled quantum states. This quantum key-distribution system proposed by Ekert is called E91 protocol.

In the E91 protocol, the transmitter Alice and the receiver Bob measure the spin of a pair coupled to total spin 0, i.e., an EPR pair. By using spin 1/2 particles, an EPR pair is expressed as

$$|\Psi\rangle_{12} = \frac{1}{\sqrt{2}}\{|\uparrow\rangle_1|\downarrow\rangle_2 - |\downarrow\rangle_1|\uparrow\rangle_2\} \tag{9.56}$$

in the same way as Eq. (4.38). An EPR pair can also be implemented by two photons, as shown by Eq. (4.67). The particles constituting the EPR pair should be emitted to the opposite directions with respect to the $z$-axis. Alice and Bob at different locations receive different particles respectively, and and they measure the direction of spin with polarization detectors. Let us use the unit vectors $\mathbf{a_i}$ and $\mathbf{b_i}$ ($i = 1, 2, 3$) to express the directions of the detectors, and assume that $\mathbf{a_i}$ and $\mathbf{b_i}$ are put in the $xy$-plane perpendicular to the $z$-axis. Alice uses $\mathbf{a_i}$ while Bob uses $\mathbf{b_j}$ to measure the direction of spin. The correlation between upward + and downward −is calculated by the correlation function (4.46) as the scalar product,

$$\begin{aligned} E(\mathbf{a_i}, \mathbf{b_j}) &= E_{++}(\mathbf{a_i}, \mathbf{b_j}) + E_{--}(\mathbf{a_i}, \mathbf{b_j}) - E_{+-}(\mathbf{a_i}, \mathbf{b_j}) - E_{-+}(\mathbf{a_i}, \mathbf{b_j}) \\ &= -(\mathbf{a_i}, \mathbf{b_j}) \end{aligned} \tag{9.57}$$

of vectors $\mathbf{a_i}$ and $\mathbf{b_j}$, where we have omitted the symbol "QM" of Eq. (4.46). In Eq. (9.57), for example, $E_{++}(\mathbf{a_i}, \mathbf{b_j})$ means the observation of spin direction with $\mathbf{a_i}$ being + and that with $\mathbf{b_j}$ being +. As shown in Fig. 9.5, the detectors with Alice and Bob are arranged at angles $(\mathbf{a_1}, \mathbf{a_2}, \mathbf{a_3}) = (\phi_1 = 0, \phi_2 = \frac{\pi}{4}, \phi_3 = \frac{\pi}{2})$ and $(\mathbf{b_1}, \mathbf{b_2}, \mathbf{b_3}) = (\theta_1 = \frac{\pi}{4}, \theta_2 = \frac{\pi}{2}, \theta_3 = \frac{3}{4}\pi)$, respectively, from the $x$-axis. Let us note that the detectors $\mathbf{a_2}$ and $\mathbf{b_1}$ have the same direction as well as $\mathbf{a_3}$ and $\mathbf{b_2}$. If these two detectors measure an EPR pair, the correlation coefficient $P$ corresponds to perfect anti-correlation, since the observation should be either $E_{+-}$ or $E_{-+}$;

$$E(\mathbf{a_2}, \mathbf{b_1}) = -\cos(\phi_2 - \theta_1) = -\cos\left(\frac{\pi}{4} - \frac{\pi}{4}\right) = -1 \tag{9.58}$$

[4]A. K. Ekert, Phys. Rev. Lett. **67**, 661 (1991).

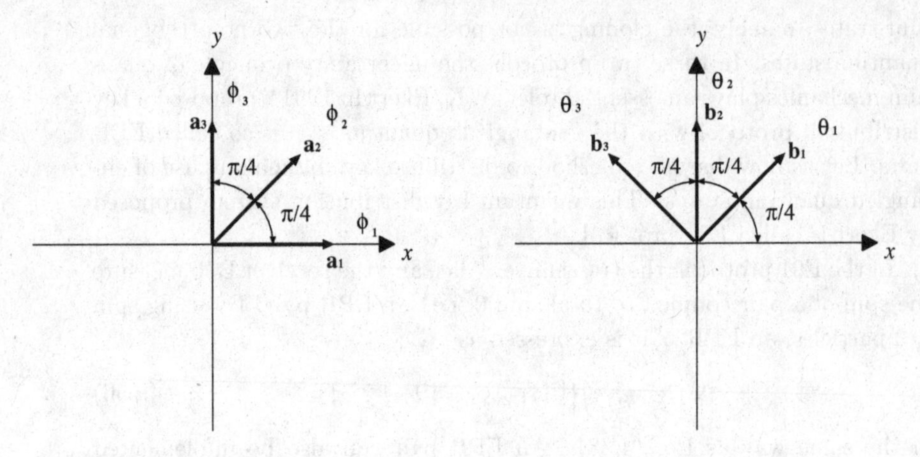


Fig. 9.5 The directions of the detectors with Alice and Bob. Alice and Bob have three detectors, respectively. In the E91 protocol, the detectors $\mathbf{a_1}$, $\mathbf{a_2}$ and $\mathbf{a_3}$ are arranged at the angles $\phi_1 = 0$, $\phi_2 = \frac{\pi}{4}$ and $\phi_3 = \frac{\pi}{2}$ from the $x$-axis. On the other hand, the detectors $\mathbf{b_1}$, $\mathbf{b_2}$ and $\mathbf{b_3}$ are arranged at the angles $\theta_1 = \frac{\pi}{4}$, $\theta_2 = \frac{\pi}{2}$ and $\theta_3 = \frac{3}{4}\pi$ from the $x$-axis.

$$E(\mathbf{a_3}, \mathbf{b_2}) = -\cos(\phi_3 - \theta_2) = -\cos\left(\frac{\pi}{2} - \frac{\pi}{2}\right) = -1. \tag{9.59}$$

Now, let us define the **CHSH inequality** (Clauser, Horne, Shimony, Holt inequality), which is a variation of Bell's inequality, as

$$S = E(\mathbf{a_1}, \mathbf{b_1}) - E(\mathbf{a_1}, \mathbf{b_3}) + E(\mathbf{a_3}, \mathbf{b_1}) + E(\mathbf{a_3}, \mathbf{b_3}). \tag{9.60}$$

Equation (9.60) means that Alice and Bob must perform measurement on the EPR pair with four different sets of the specific directions. By substituting the result of Eq. (4.46) of Chapter 4 into Eq. (9.60), we obtain that the results of quantum-mechanical measurements as

$$S_{QM} = -\cos(\phi_1 - \theta_1) + \cos(\phi_1 - \theta_3) - \cos(\phi_3 - \theta_1) - \cos(\phi_3 - \theta_3). \tag{9.61}$$

For the detector arrangements in Fig. 9.5, we have

$$S_{QM} = -\cos\left(-\frac{\pi}{4}\right) + \cos\left(-\frac{3}{4}\pi\right) - \cos\left(\frac{\pi}{4}\right) - \cos\left(-\frac{\pi}{4}\right) = -2\sqrt{2}. \tag{9.62}$$

As a matter of fact, the result (9.62) gives the maximum anti-correlation of measurement along four different directions specified by angles $\phi_i$ and $\theta_i$ (See Exercise [2] in this chapter).

**E91 protocol**

**Step 1** :

Alice and Bob perform measurements with their own detectors arranged as shown in Fig. 9.5.

**Step 2** :

Alice and Bob make public the direction of the detectors. They classify the results of measurements into two groups:

(a) results by detectors at different directions,

(b) results by detectors at the same directions.

**Step 3** :

Alice and Bob make public only those results in group (a), and check whether the CHSH inequality $S$ is satisfied.

**Step 4** :

They adopt the results of group (b) as the one-time pad, i.e., as the secret key only if the CHSH inequality $S = -2\sqrt{2}$ is satisfied.

Assume that Alice and Bob distribute the secret key after the results of measurements on EPR pairs by the E91 protocol described above. The problem will be how the CHSH changes if a tapper tries to steal the secret key by breaking the E91 protocol. Let us derive the CHSH inequality of the case where the tapper "Eve" tries to steal the information associated with the EPR pair. The first characteristic of the E91 protocol lies in the point that Eve cannot acquire any information even if she measures all the EPR pairs that Alice and Bob are trying to measure. That is because the secret key is not generated by the EPR pairs themselves but is generated by choosing only the data set of Step 2 (b) after the measurements by Alice and Bob. If the tapper Eve gets information about the cipher key, she needs to send EPR pairs again to Alice and Bob after her own measurements on the EPR pairs, and to know the types of detectors with Alice and Bob. To be more specific, let us consider the case where Eve measure the spins of the EPR pairs along the directions $\phi_A$ and $\theta_B$ using two detectors $A$ and $B$, and then send a pair of particles of the spins of the same directions as she observes. If Eve adopts the directions $\phi_A$ and $\theta_B$ with certain probabilities $p(\phi_A, \theta_B) = |a(\phi_A, \theta_B)|^2$, the state of the two delivered particles is expressed, using the amplitude $a(\phi_A, \theta_B)$, as

$$|\Psi(A,B)\rangle = \int_0^{2\pi} d\phi_A \int_0^{2\pi} d\theta_B \, a(\phi_A, \theta_B)|\phi_A\rangle_1 |\theta_B\rangle_2 \,, \qquad (9.63)$$

where, since the state $|\phi_A\rangle$, $|\theta_B\rangle$ is normalized as $\langle\phi_A|\phi'_A\rangle = \delta(\phi_A - \phi'_A)$,

$\langle\theta_B|\theta'_B\rangle = \delta(\theta_B - \theta'_B)$, the state $|\Psi(A,B)\rangle$ is also normalized as

$$\langle\Psi(A,B)|\Psi(A,B)\rangle = \int_0^{2\pi}\int_0^{2\pi} |a(\phi_A,\theta_B)|^2 d\phi_A d\theta_B$$
$$= \int_0^{2\pi}\int_0^{2\pi} p(\phi_A,\theta_B) d\phi_A d\theta_B = 1. \tag{9.64}$$

When Alice and Bob measure the direction of the spin of the state $|\phi_A\rangle|\theta_B\rangle$ sent by Eve with the detectors **a** and **b**, the result $E'$ in the presence of a tapper is derived by Eq. (4.46) as

$$E'(\mathbf{a},\mathbf{b}) = \int\int p(\phi_A,\theta_B)_1\langle\phi_A|_2\langle\theta_B|(\boldsymbol{\sigma}_1\cdot\mathbf{a})(\boldsymbol{\sigma}_2\cdot\mathbf{b})|\phi_A\rangle_1|\theta_B\rangle_2 d\phi_A d\theta_B$$
$$= \int\int p(\phi_A,\theta_B)(\mathbf{n_A}\cdot\mathbf{a})(\mathbf{n_B}\cdot\mathbf{b}) d\phi_A d\theta_B, \tag{9.65}$$

where $\mathbf{n_A}$ and $\mathbf{n_B}$ are the unit vectors of the same directions as the spins of the states $|\phi_A\rangle$ and $|\theta_B\rangle$, and the angles between the $x$-axis and two spins are $\phi_A$ and $\theta_B$, respectively. Letting $\phi$ be the angle between the $x$-axis and **a**, and $\theta$ between the $x$-axis and **b**, we can express $E'$ as

$$E'(\mathbf{a},\mathbf{b}) = \int\int p(\phi_A,\theta_B)\cos(\phi-\phi_A)\cos(\theta-\theta_B) d\phi_A d\theta_B. \tag{9.66}$$

If Alice and Bob arrange the detectors $\mathbf{a_i}, \mathbf{b_i}$ $(i = 1,2,3)$ as in Fig. 9.5, the correlation coefficient $S'$ is given by

$$S' = E'(\mathbf{a_1},\mathbf{b_1}) - E'(\mathbf{a_1},\mathbf{b_3}) + E'(\mathbf{a_3},\mathbf{b_1}) + E'(\mathbf{a_3},\mathbf{b_3})$$
$$= \int\int p(\phi_A,\theta_B)\left[\cos(-\phi_A)\cos\left(\frac{\pi}{4}-\theta_B\right) - \cos(-\phi_A)\cos\left(\frac{3}{4}\pi-\theta_B\right)\right.$$
$$\left. + \cos\left(\frac{\pi}{2}-\phi_A\right)\cos\left(\frac{\pi}{4}-\theta_B\right) + \cos\left(\frac{\pi}{2}-\phi_A\right)\cos\left(\frac{3}{4}\pi-\theta_B\right)\right] d\phi_A d\theta_B$$
$$= \int\int p(\phi_A,\theta_B)\sqrt{2}\cos(\phi_A-\theta_B) d\phi_A d\theta_B. \tag{9.67}$$

Using the normalization condition (9.64), the absolute value of $S'$ in Eq. (9.67) becomes

$$|S'| = \int\int p(\phi_A,\theta_B)\sqrt{2}\,|\cos(\phi_A-\theta_B)|\, d\phi_A d\theta_B$$
$$\leq \int\int p(\phi_A,\theta_B) d\phi_A d\theta_B \sqrt{2} = \sqrt{2}. \tag{9.68}$$

As a result of Eve's tapping, the correlation coefficient becomes

$$-\sqrt{2} \leq S' \leq \sqrt{2}, \tag{9.69}$$

which is in contradiction with the quantum mechanical result of Eq. (9.62), $S_{QM} = -2\sqrt{2}$. In this way, Alice and Bob can recognize the presence of a tapper by making use the CHSH inequality, i.e., the generalized Bell theorem.

The measurement of the EPR pair by the tapper results in determining the direction of the spin. In other words, the quantum-mechanical two particle state is contracted by the measurement, and the directions of the spins are fixed. This physical state corresponds to the state given as Eq. (4.10) by the hidden variable theory of Einstein in Chapter 4. Specifically, by the presence of the tapper "Eve", the EPR pair is disturbed, and the classical Bell inequality holds. The capability of recognizing tappers by the measurement at a certain number of different directions is a wonderful application of the principles of quantum mechanics with the entangled states.

**Summary of Chapter 9**

(1) The cryptography systems with secret keys have been used from ancient days. But they were often broken when the key was stolen.
(2) The public key cryptography system was invented as a safer communication. Even if the public key for encoding is stolen, the risk of deciphering is very low because a different secret key is needed for decoding.
(3) Quantum algorithms can be used to solve the problem of obtaining the secret key from the public key.
(4) Various key distribution systems (BB84, B92, E91) using qubits are proposed. The quantum disturbance by measurement is utilized to find a tapper based on the non-cloning theorem.

## 9.5 Problems

[1] Derive

$$\cos(-\phi_A)\cos\left(\frac{\pi}{4}-\theta_B\right) - \cos(-\phi_A)\cos\left(\frac{3}{4}\pi-\theta_B\right) + \cos\left(\frac{\pi}{2}-\phi_A\right)\cos\left(\frac{\pi}{4}-\theta_B\right) + \cos\left(\frac{\pi}{2}-\phi_A\right)\cos\left(\frac{3}{4}\pi-\theta_B\right) = \sqrt{2}\cos(\phi_A-\theta_B) \tag{9.70}$$

in Eq. (9.67).

[2] "Proof of the fact that Eq. (9.62) is the maximum anti-correlation in the observation of an EPR pair." As shown in Fig. 9.6, Choose the

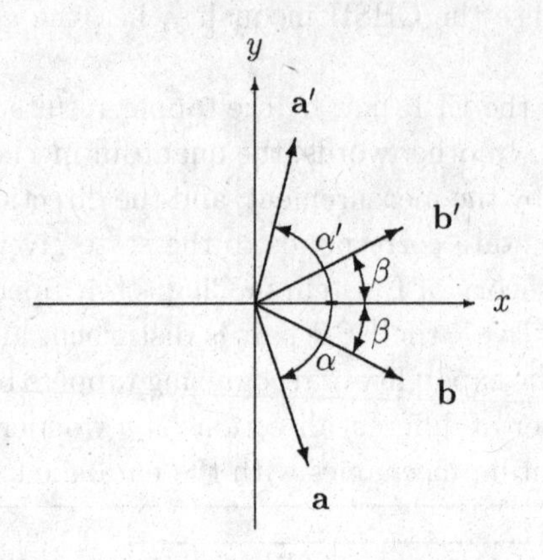


Fig. 9.6 Measurement of an EPR pair.

coordinates so that the $x$-axis bisects $\mathbf{b}$ and $\mathbf{b}'$. Let $\beta$ the common angle between the two vectors and the $x$-axis. $(0 \leq \beta \leq \pi/2)$.

(a) Let $\alpha$ the angle between $\mathbf{a}$ and the $x$-axis $(-\pi < \alpha \leq \pi)$. Fix $\beta$, and vary only $\alpha$. Obtain the maximum and the minimum of

$$|P(\mathbf{a}, \mathbf{b}) - P(\mathbf{a}, \mathbf{b}')|$$

as a function of $\beta$.

(b) Let $\alpha'$ the angle between $\mathbf{a}'$ and the $x$-axis $(-\pi < \alpha' \leq \pi)$. Fix $\beta$, and vary only $\alpha'$. Obtain the maximum and the minimum of

$$|P(\mathbf{a}', \mathbf{b}) + P(\mathbf{a}', \mathbf{b}')|$$

as a function of $\beta$.

(c) Varying $\alpha$, $\alpha'$ and $\beta$, obtain the maximum and the minimum of $|S_{\text{QM}}|$.

## Chapter 10

# Quantum Search Algorithm

The quantum search algorithm was discovered by L. K. Grover in 1997, and is called Grover's algorithm after his name.[1] Let us consider how to find elements marked by a certain symbol out of a database consisting of $N$ elements (say, multi-million elements). This problem is similar to how to find the name of the person from a telephone book knowing only his/her telephone number. This search takes by far more time than finding the phone number from the name of the person. The difference comes from the structure of the telephone book. In the telephone book, the names are arranged in the alphabetical order and constitute the structured database, while phone numbers do not form any structure in the telephone book. In classical computers, when we search one target element from the database with $N$ elements, we need $N/2$ trials on average. Grover showed using the quantum search algorithm that $\sqrt{N}$ trials will be enough on average to find the target element from the database with $N$ elements. Shor's quantum computation algorithm was an epoch-making algorithm which made possible to solve the factorization problem in polynomial time , while the classical computers need exponentially increasing time to solve it. In comparison, Grover's algorithm makes the data search possible by $O(\sqrt{N})$ computation time, while it takes $O(N)$ computation time by the classical computers. Concerning the reduction of computing time, Grover's algorithm may look less powerful than Shor's algorithm. However, it has good potentiality of application in the wide area such as NP-complete problems and structured databases.

**Keywords: oracle function (black box); quantum oracle; Grover's search algorithm.**

[1] L. K. Grover, Phys. Rev. Lett. **79**, 325 (1997).

## 10.1 Oracle function

In Grover's algorithm, the concept of **oracle**, or black box, is introduced. Let us think of the problem of searching an element labeled $z_0$ out of the database with $N = 2^n$ elements. We introduce a binary function mapping a set of $2^n$ integers $x \in \{0, 1, 2, \cdots, 2^n - 1\}$ on a set of two integers $\{0, 1\}$:

$$f: \{0,1\}^n \longrightarrow \{0,1\}, \tag{10.1}$$

which is called "oracle" or "black box". In the present problem, we define $f(x)$ as

$$f(x) = \begin{cases} 1 \text{ if } x = z_0, \\ 0 \text{ otherwise.} \end{cases} \tag{10.2}$$

Note that a similar function was introduced in the algorithm of Deutsch and Jozsa in Section 8.2. We assume that the oracle function can be called any time as a subroutine, and the computation time is proportional to the number of calls. In classical computers, $N/2$ calls are needed on average to find the solution $z_0$; namely, the computation time of $O(N) = O(2^n)$ is needed.

## 10.2 Quantum oracle

We introduce the quantum oracle, or quantum black box which acts as the oracle function $f$. The oracle was also used in the quantum computation by Deutsch and Jozsa algorithm. The quantum oracle is defined by a unitary operator $U_f$ as

$$U_f|x\rangle|q\rangle = |x\rangle|q \oplus f(x)\rangle. \tag{10.3}$$

The first register $|x\rangle$ is an index register with $n$ bits, while the second register is called the oracle bit, where the symbol $\oplus$ means the logical operation of exclusive OR; namely, it gives the sum of $q$ and $f(x)$ in a congruence equation modulo 2,

$$0 \oplus 0 = 0, \qquad 1 \oplus 0 = 1, \qquad 0 \oplus 1 = 1, \qquad 1 \oplus 1 = 0. \tag{10.4}$$

From Eq. (10.3), one can decide whether $x$ is the solution or not measuring the second register; Initially, the second register is set to be $q = 0$ and applied the oracle operator $U_f$. If the second register remains 0, $x$ is not the solution. If the second register has turned to be 1, the value in the first register gives the solution $z_0$. Thus the quantum oracle of Eq. (10.3) has the effect equivalent to the classical one.

The characteristic of the quantum search algorithm is based on the choice of the oracle bit as a superposed state, $\frac{1}{\sqrt{2}}(|0\rangle - |1\rangle)$. For this choice, we have

$$U_f|x\rangle\frac{1}{\sqrt{2}}(|0\rangle - |1\rangle) = \begin{cases} |x\rangle\frac{1}{\sqrt{2}}(|1\rangle - |0\rangle) & x = z_0 \\ |x\rangle\frac{1}{\sqrt{2}}(|0\rangle - |1\rangle) & x \neq z_0 \end{cases}$$

$$= (-1)^{f(x)}|x\rangle\frac{1}{\sqrt{2}}(|0\rangle - |1\rangle), \tag{10.5}$$

namely, the quantum oracle operator gives the phase $(-1)^{f(x)}$. By ignoring the second register, we can write the result as

$$U_f|x\rangle = \begin{cases} -|x\rangle & x = z_0, \\ |x\rangle & x \neq z_0. \end{cases} \tag{10.6}$$

Thus the solution can be found by the quantum oracle in terms of the phase change of the state $|x\rangle$. If we take into account only the index bit, we can express the equivalent operator to $U_f$ as

$$U(z_0) = \mathrm{I} - 2|z_0\rangle\langle z_0|, \tag{10.7}$$

where I stands for the identity transformation, i.e., the unit matrix. Using $\langle x|z_0\rangle = \delta_{x,z_0}$, it is quite easy to verify that $U(z_0)|x\rangle$ gives (10.6).

In attaining Grover's algorithm, we need another unitary operator

$$U(0) = \mathrm{I} - 2|0\rangle\langle 0| \tag{10.8}$$

in addition to (10.7). The unitary operator $U(0)$ of Eq. (10.8) gives the phase $(-1)$ only to the state $|0\rangle$ as

$$U(0)|x\rangle = (-1)^{\delta_{x,0}}|x\rangle. \tag{10.9}$$

By applying the Hadamard transformation $H^{\otimes n}$ to both sides of the operator (10.8), we have

$$U(\psi) = H^{\otimes n}(\mathrm{I} - 2|0\rangle\langle 0|)H^{\otimes n} = \mathrm{I} - 2|\psi\rangle\langle\psi|, \tag{10.10}$$

where the state $|\psi\rangle$ is defined as

$$|\psi\rangle = H^{\otimes n}|0\rangle = \frac{1}{\sqrt{2^n}}\sum_{x=0}^{N-1}|x\rangle. \tag{10.11}$$

In deriving (10.10), the relation $H^{\otimes n}\mathrm{I}H^{\otimes n} = H^{2\otimes n} = \mathrm{I}$ has been used. With $U(z_0)$ and $U(\psi)$, Grover's operator is defined as

$$U(G) = -U(\psi)U(z_0). \tag{10.12}$$

The operator (10.12) is also called Grover's iteration operator, which is applied repeatedly until the solution is obtained in the index register. Let us see that the operator (10.12) works as a rotation operator in the two-dimensional space spanned by two components of the state vector $|\psi\rangle$, i.e., one is the state vector of the solution, and another is composed of the state vectors of the non-solutions. In general cases, we assume that there are $M$ solutions $|z_i\rangle$ in $|\psi\rangle$. Then, we define the superposition of the solution states,

$$|\beta\rangle = \frac{1}{\sqrt{M}} \sum_{z_i=0}^{M-1} |z_i\rangle, \tag{10.13}$$

and the $(N - M)$ non-solutions states $|y_i\rangle$ as

$$|\alpha\rangle = \frac{1}{\sqrt{N-M}} \sum_{y_i=0}^{N-1} |y_i\rangle. \tag{10.14}$$

Then the initial state $|\psi\rangle$ can be written as

$$|\psi\rangle = \sqrt{\frac{N-M}{N}}|\alpha\rangle + \sqrt{\frac{M}{N}}|\beta\rangle. \tag{10.15}$$

As shown in Fig. 10.1, in the two-dimensional space $(\alpha, \beta)$ with the abscissa $|\alpha\rangle$ and the ordinate $|\beta\rangle$, the state $|\psi\rangle$ is considered as a unit vector having an angle $\theta/2$ with the $\alpha$-axis. The angle $\theta/2$ is given by

$$\cos\frac{\theta}{2} = \sqrt{\frac{N-M}{N}}, \quad \sin\frac{\theta}{2} = \sqrt{\frac{M}{N}}. \tag{10.16}$$

Using Eq. (10.16), we can express the state $\psi$ of Eq. (10.15) as

$$|\psi\rangle = \cos\frac{\theta}{2}|\alpha\rangle + \sin\frac{\theta}{2}|\beta\rangle. \tag{10.17}$$

In the case with $M$ solutions, the operator $U(z_0)$ in Eq. (10.7) is replaced by

$$U(\beta) = \mathrm{I} - 2|\beta\rangle\langle\beta|. \tag{10.18}$$

In the two-dimensional space of $(\alpha, \beta)$, the operator $U(\beta)$ can be represented by the following $2 \times 2$ matrix,

$$U(\beta) = \begin{matrix} & |\alpha\rangle \quad |\beta\rangle & \\ & \begin{pmatrix} 1 & 0 \\ 0 & -1 \end{pmatrix} & \begin{matrix} |\alpha\rangle \\ |\beta\rangle \end{matrix} \end{matrix}. \tag{10.19}$$

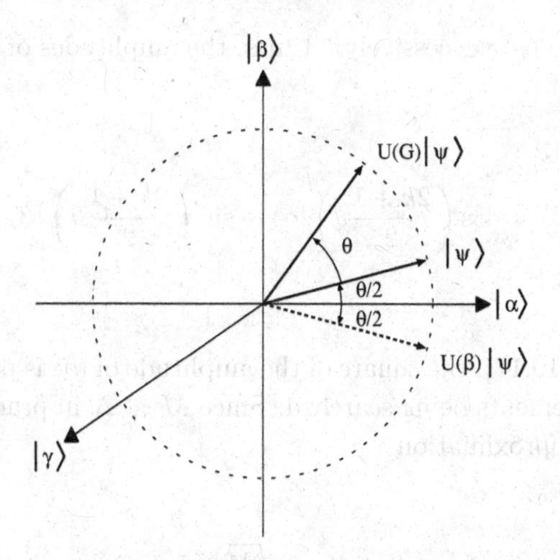


Fig. 10.1 Action of $U(\beta)$ ($U(z_0)$ in the case of one solution) and $U(G)$ upon $|\psi\rangle$. The $\alpha$-axis, the $\beta$-axis and $\gamma$-axis compose a three-dimensional rectilinear coordinates. The operator $U(\beta)$ acts as a reflection operator on the vector $|\psi\rangle$ with respect to the $(\alpha,\gamma)$-plane, while $U(G)$ rotates the vector $|\psi\rangle$ by an angle $\theta$ about the $\gamma$-axis.

As shown in Fig. 10.1, this operator acts as a reflection operator for the vector $|\psi\rangle$ on the $(\alpha\gamma)$-plane. On the other hand, $-U(\psi)$ can be expressed as

$$-U(\psi) = 2|\psi\rangle\langle\psi| - \mathrm{I} = 2\begin{pmatrix} \cos^2\frac{\theta}{2} & \cos\frac{\theta}{2}\sin\frac{\theta}{2} \\ \cos\frac{\theta}{2}\sin\frac{\theta}{2} & \sin^2\frac{\theta}{2} \end{pmatrix} - \begin{pmatrix} 1 & 0 \\ 0 & 1 \end{pmatrix}$$
$$= \begin{pmatrix} \cos\theta & \sin\theta \\ \sin\theta & -\cos\theta \end{pmatrix}. \tag{10.20}$$

From Eq. (10.19) and Eq. (10.20), we have

$$U(G) = -U(\psi)U(\beta) = \begin{pmatrix} \cos\theta & \sin\theta \\ \sin\theta & -\cos\theta \end{pmatrix}\begin{pmatrix} 1 & 0 \\ 0 & -1 \end{pmatrix} = \begin{pmatrix} \cos\theta & -\sin\theta \\ \sin\theta & \cos\theta \end{pmatrix}. \tag{10.21}$$

The matrix representation (10.21) of the operator $U(G)$ shows that $U(G)$ rotates the vector by an angle $\theta$ about the $\gamma-$axis which is perpendicular to the $(\alpha,\beta)$ plane. Thus, $U(\beta)$ acts as the reflection operator with respect to $(\alpha,\beta)$, and $U(G)$ is the rotation operator about the $\gamma-$axis. In this way, we find that $U(G)$ has a definite geometrical meaning, and rotates the state vector $|\psi\rangle$ by $\theta$ in the $(\alpha,\beta)$-plane as

$$|\psi\rangle_1 = U(G)|\psi\rangle = \begin{pmatrix} \cos\theta & -\sin\theta \\ \sin\theta & \cos\theta \end{pmatrix}\begin{pmatrix} \cos\frac{\theta}{2} \\ \sin\frac{\theta}{2} \end{pmatrix} = \begin{pmatrix} \cos\frac{3}{2}\theta \\ \sin\frac{3}{2}\theta \end{pmatrix}$$
$$= \cos\frac{3}{2}\theta|\alpha\rangle + \sin\frac{3}{2}\theta|\beta\rangle. \tag{10.22}$$

In applying $U(G)$ to $|\psi\rangle$ successively $k$ times, the amplitudes of $|\alpha\rangle$ and $|\beta\rangle$ vary as

$$|\psi\rangle_k = U(G)^k|\psi\rangle = \cos\left(\frac{2k+1}{2}\theta\right)|\alpha\rangle + \sin\left(\frac{2k+1}{2}\theta\right)|\beta\rangle. \tag{10.23}$$

In the initial state (10.17), the square of the amplitude of $|\beta\rangle$ is proportional to the number of elements being searched. Since $M \ll N$ in practical cases, we can make the approximation

$$\sin\frac{\theta}{2} \approx \frac{\theta}{2} = \sqrt{\frac{M}{N}}. \tag{10.24}$$

As we apply successively Grover's operators $U(G)$, the amplitude of the searched state $|\beta\rangle$ is continuously growing. When the angle approaches $\pi/2$, the probability of finding the solution gets higher. Finally, the solution is obtained if the condition

$$k\theta \approx 2k\sqrt{\frac{M}{N}} \approx \frac{\pi}{2} \tag{10.25}$$

is satisfied. The number of applications for the condition is estimated as

$$k \approx \frac{\pi}{4}\sqrt{\frac{N}{M}} \approx O(\sqrt{N}), \tag{10.26}$$

which is proportional to the square root of the number of data $N$. By classical computers, the search needs $O(N)$ operations. The remarkable point of Grover's algorithm is that it can finish the search by $O(\sqrt{N})$ operations.

The quantum search algorithm is summarized as follows:

**Quantum Search Algorithm (Grover's Search Algorithm)**

Algorithm to find the data with index $z_0$ out of a database with $N = 2^n$ elements:

**Step 1**: Prepare a state with index $n$ bits and an oracle bit:

$$|\psi_1\rangle = |0\rangle^{\otimes n}|0\rangle \tag{10.27}$$

**Step 2**: Apply the Hadamard transformation $H^{\otimes n}$ to the $n$ bits, and apply $HX$ to the oracle bit, where $X$ means the negation gate.

$$|\psi_2\rangle = H^{\otimes n}HX|\psi_1\rangle = \frac{1}{\sqrt{2^n}}\sum_{x=0}^{N-1}|x\rangle\left[\frac{|0\rangle - |1\rangle}{\sqrt{2}}\right] \tag{10.28}$$

**Step 3**: Apply Grover's operator $U(G) = -U(\psi)U(\beta)$ by $k \approx \pi\sqrt{N}/4$ times.

$$|\psi_3\rangle = U(G)^k|\psi_2\rangle \approx |z_0\rangle\left[\frac{|0\rangle - |1\rangle}{\sqrt{2}}\right] \tag{10.29}$$

**Step 4**: The solution $|z_0\rangle$ is obtained by observing the index $n$ bit state.

---

Note: The oracle operator $U(\beta)$ ($U(z_0)$ in the case of single solution) is defined as

$$U(\beta)|x\rangle|q\rangle = |x\rangle|q \oplus f(x)\rangle, \tag{10.30}$$

where

$$f(x) = \begin{cases} 0 & x \neq z_0, \\ 1 & x = z_0. \end{cases} \tag{10.31}$$

The symbol $\oplus$ stands for the exclusive OR in the binary system. The operator $U(\psi)$ is defined as

$$U(\psi) = H^{\otimes n}\{I - 2|0\rangle\langle 0|\}H^{\otimes n}. \tag{10.32}$$

---

**Example 8.1** Using the quantum search algorithm, find an unknown element with label $z_0 = 2$ out of a database with $N = 2^3$ elements.

**Solution** Let us express Grover's operator in the form of an $8 \times 8$

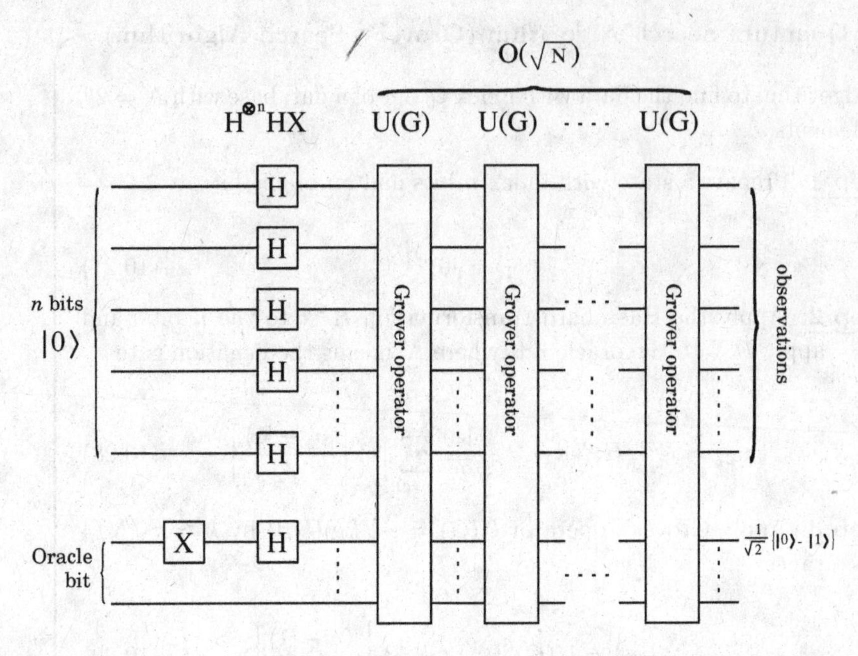


Fig. 10.2 Circuit of quantum search algorithm. The symbol $H$ stands for the Hadamard transformation, $X$ for the negation gate, and $U(G)$ for Grover's operator. The operator $U(G)$ is applied $\sqrt{N}$ times. The size of the database is given by $N = 2^n$.

matrix. The matrix for $U(z_0)$ is derived as

$$U(z_0) = \mathrm{I} - 2|z_0 = 2\rangle\langle z_0 = 2|$$

$$= \begin{pmatrix} 1&0&0&0&0&0&0&0 \\ 0&1&0&0&0&0&0&0 \\ 0&0&1&0&0&0&0&0 \\ 0&0&0&1&0&0&0&0 \\ 0&0&0&0&1&0&0&0 \\ 0&0&0&0&0&1&0&0 \\ 0&0&0&0&0&0&1&0 \\ 0&0&0&0&0&0&0&1 \end{pmatrix} - 2 \begin{pmatrix} 0&0&0&0&0&0&0&0 \\ 0&0&0&0&0&0&0&0 \\ 0&0&1&0&0&0&0&0 \\ 0&0&0&0&0&0&0&0 \\ 0&0&0&0&0&0&0&0 \\ 0&0&0&0&0&0&0&0 \\ 0&0&0&0&0&0&0&0 \\ 0&0&0&0&0&0&0&0 \end{pmatrix}$$

$$= \begin{pmatrix} 1&0&0&0&0&0&0&0 \\ 0&1&0&0&0&0&0&0 \\ 0&0&-1&0&0&0&0&0 \\ 0&0&0&1&0&0&0&0 \\ 0&0&0&0&1&0&0&0 \\ 0&0&0&0&0&1&0&0 \\ 0&0&0&0&0&0&1&0 \\ 0&0&0&0&0&0&0&1 \end{pmatrix}. \tag{10.33}$$

The matrix for $U(\psi)$ is obtained, using (10.10) and (10.11), as

$$U(\psi) = \mathrm{I} - 2|\psi\rangle\langle\psi|$$

$$= \begin{pmatrix} 1 & 0 & 0 & 0 & 0 & 0 & 0 & 0 \\ 0 & 1 & 0 & 0 & 0 & 0 & 0 & 0 \\ 0 & 0 & 1 & 0 & 0 & 0 & 0 & 0 \\ 0 & 0 & 0 & 1 & 0 & 0 & 0 & 0 \\ 0 & 0 & 0 & 0 & 1 & 0 & 0 & 0 \\ 0 & 0 & 0 & 0 & 0 & 1 & 0 & 0 \\ 0 & 0 & 0 & 0 & 0 & 0 & 1 & 0 \\ 0 & 0 & 0 & 0 & 0 & 0 & 0 & 1 \end{pmatrix} - 2\frac{1}{8} \begin{pmatrix} 1 & 1 & 1 & 1 & 1 & 1 & 1 & 1 \\ 1 & 1 & 1 & 1 & 1 & 1 & 1 & 1 \\ 1 & 1 & 1 & 1 & 1 & 1 & 1 & 1 \\ 1 & 1 & 1 & 1 & 1 & 1 & 1 & 1 \\ 1 & 1 & 1 & 1 & 1 & 1 & 1 & 1 \\ 1 & 1 & 1 & 1 & 1 & 1 & 1 & 1 \\ 1 & 1 & 1 & 1 & 1 & 1 & 1 & 1 \\ 1 & 1 & 1 & 1 & 1 & 1 & 1 & 1 \end{pmatrix}$$

$$= \frac{1}{4} \begin{pmatrix} 3 & -1 & -1 & -1 & -1 & -1 & -1 & -1 \\ -1 & 3 & -1 & -1 & -1 & -1 & -1 & -1 \\ -1 & -1 & 3 & -1 & -1 & -1 & -1 & -1 \\ -1 & -1 & -1 & 3 & -1 & -1 & -1 & -1 \\ -1 & -1 & -1 & -1 & 3 & -1 & -1 & -1 \\ -1 & -1 & -1 & -1 & -1 & 3 & -1 & -1 \\ -1 & -1 & -1 & -1 & -1 & -1 & 3 & -1 \\ -1 & -1 & -1 & -1 & -1 & -1 & -1 & 3 \end{pmatrix}. \tag{10.34}$$

Therefore, Grover's operator is given by

$$U(G) = -U(\psi)U(z_0)$$

$$= -\frac{1}{4} \begin{pmatrix} 3 & -1 & -1 & -1 & -1 & -1 & -1 & -1 \\ -1 & 3 & -1 & -1 & -1 & -1 & -1 & -1 \\ -1 & -1 & 3 & -1 & -1 & -1 & -1 & -1 \\ -1 & -1 & -1 & 3 & -1 & -1 & -1 & -1 \\ -1 & -1 & -1 & -1 & 3 & -1 & -1 & -1 \\ -1 & -1 & -1 & -1 & -1 & 3 & -1 & -1 \\ -1 & -1 & -1 & -1 & -1 & -1 & 3 & -1 \\ -1 & -1 & -1 & -1 & -1 & -1 & -1 & 3 \end{pmatrix} \begin{pmatrix} 1 & 0 & 0 & 0 & 0 & 0 & 0 & 0 \\ 0 & 1 & 0 & 0 & 0 & 0 & 0 & 0 \\ 0 & 0 & -1 & 0 & 0 & 0 & 0 & 0 \\ 0 & 0 & 0 & 1 & 0 & 0 & 0 & 0 \\ 0 & 0 & 0 & 0 & 1 & 0 & 0 & 0 \\ 0 & 0 & 0 & 0 & 0 & 1 & 0 & 0 \\ 0 & 0 & 0 & 0 & 0 & 0 & 1 & 0 \\ 0 & 0 & 0 & 0 & 0 & 0 & 0 & 1 \end{pmatrix}$$

$$= \frac{1}{4} \begin{pmatrix} -3 & 1 & -1 & 1 & 1 & 1 & 1 & 1 \\ 1 & -3 & -1 & 1 & 1 & 1 & 1 & 1 \\ 1 & 1 & 3 & 1 & 1 & 1 & 1 & 1 \\ 1 & 1 & -1 & -3 & 1 & 1 & 1 & 1 \\ 1 & 1 & -1 & 1 & -3 & 1 & 1 & 1 \\ 1 & 1 & -1 & 1 & 1 & -3 & 1 & 1 \\ 1 & 1 & -1 & 1 & 1 & 1 & -3 & 1 \\ 1 & 1 & -1 & 1 & 1 & 1 & 1 & -3 \end{pmatrix}. \tag{10.35}$$

Note that the third column has a phase different from others. It will be made clear in Step 3 that this phase is the key point of the data search by Grover's algorithm.

**Step 1**:

Prepare a database with 3 bits and an oracle bit with 1 bit as the initial state:

$$|\psi_1\rangle = |0\rangle^{\otimes 3}|0\rangle. \tag{10.36}$$

**Step 2**:

Apply the Hadamard transformation $H^{\otimes 3}$ to the index 3 bits, and $HX$ to the oracle bit:

$$|\psi_2\rangle = \frac{1}{\sqrt{8}}\sum_{x=0}^{7}|x\rangle\left[\frac{|0\rangle - |1\rangle}{\sqrt{2}}\right]. \tag{10.37}$$

**Step 3**:

Apply Grover's operator $U(G)$ by $k = \pi\sqrt{8}/4 \approx 2$ times. Ignoring the oracle bit, we have

$$U(G)|\psi_2\rangle = \frac{1}{4}\begin{pmatrix} -3 & 1 & -1 & 1 & 1 & 1 & 1 & 1 \\ 1 & -3 & -1 & 1 & 1 & 1 & 1 & 1 \\ 1 & 1 & 3 & 1 & 1 & 1 & 1 & 1 \\ 1 & 1 & -1 & -3 & 1 & 1 & 1 & 1 \\ 1 & 1 & -1 & 1 & -3 & 1 & 1 & 1 \\ 1 & 1 & -1 & 1 & 1 & -3 & 1 & 1 \\ 1 & 1 & -1 & 1 & 1 & 1 & -3 & 1 \\ 1 & 1 & -1 & 1 & 1 & 1 & 1 & -3 \end{pmatrix} \frac{1}{\sqrt{8}}\begin{pmatrix} 1 \\ 1 \\ 1 \\ 1 \\ 1 \\ 1 \\ 1 \\ 1 \end{pmatrix}$$

$$= \frac{1}{4\sqrt{2}}\begin{pmatrix} 1 \\ 1 \\ 5 \\ 1 \\ 1 \\ 1 \\ 1 \\ 1 \end{pmatrix}$$

$$|\psi_3\rangle = U(G)^2|\psi_2\rangle = \frac{1}{8\sqrt{2}}\begin{pmatrix} -1 \\ -1 \\ 11 \\ -1 \\ -1 \\ -1 \\ -1 \\ -1 \end{pmatrix}. \tag{10.38}$$

**Step 4**:

Measure the index bit. The probability of obtaining the correct solution, $z_0 = 2$, is 95% because

$$P(x=2) = \frac{121}{128} = 0.95 \quad (k=2),$$
$$P(x \neq 2) = \frac{7}{128} = 0.05. \tag{10.39}$$

We can see that there is appreciably high probability even by a single ($k = 1$) application of Grover's operator:

$$P(x=2) = \frac{25}{32} = 0.78 \quad (k=1). \tag{10.40}$$

Note that the results of measurement in Eqs. (10.39) and (10.40) agree with

$$P(z_0) = \sin^2\left(\frac{2k+1}{2}\theta\right) \tag{10.41}$$

derived from (10.23) setting $\sin(\theta/2) = 1/\sqrt{8}$ and $\cos(\theta/2) = \sqrt{7/8}$.

---

**Summary of Chapter 10**

(1) The oracle is introduced as an operator which can give an answer at once whether a given input is true (1) or false (0).

(2) the quantum oracle acts to find the solutions to the search algorithm by making use of an oracle qubit.

(3) Grover's quantum search algorithm finds the true elements in $O(\sqrt{N})$ out of $N$ elements, which classical methods can do only in $O(N)$.

## 10.3 Problem

**[1]** Assuming $\sin(\theta/2) = 1/\sqrt{8}$, derive the following values using $P(z)_k = \sin^2\left(\frac{2k+1}{2}\theta\right)$:

$$P(z_0)_{k=1} = \sin^2\left(\frac{3}{2}\theta\right) = \frac{25}{32}, \tag{10.42}$$

$$\text{[6pt]}P(z_0)_{k=2} = \sin^2\left(\frac{5}{2}\theta\right) = \frac{121}{128}. \tag{10.43}$$

# Chapter 11

# Physical Devices of Quantum Computers

To implement quantum computers, one needs physical systems to perform the unitary transformations such as the Hadamard transformation on qubits. Quantum devices by atoms, molecules and atomic nuclei, or those using photons are invented as quantum computers, where the systems are controlled according to the laws of quantum mechanics.

Table 11.1 Examples of quantum circuits.

| system | logical values | control |
|---|---|---|
| atoms, molecules or atomic nuclei of spin 1/2 | spin directions ($s_z = \pm 1/2$) | magnetic field (nuclear magnetic resonance : NMR) |
| ion traps | ground state and excited state of electrons or vibration of molecules | laser beams |
| quantum dots | ground state and an excited states of electrons | electromagnetic waves (microwave) |

In various designs proposed currently, atoms and atomic nuclei are used as quantum computers controlled by electromagnetic waves including laser. The measurements are performed quantum mechanically and give the output of computations. Typically, quantum systems is controlled by radiowave or laser. This process corresponds to the data input and the computation. Detecting microwave or light from the quantum systems corresponds to reading the output. We show some examples in Table 11.1. In the following, we will show how to implement specific processes of quantum computers;

- how to input data
- how to read outputs

- how to communicate between devices
- computation speed.

Since the process of reading results is a kind of measurements, it is a non-continuous process in quantum mechanics that disturbs the system. The disturbance on quantum computers by measurement and other effects is called decoherence. **The decoherence time** is determined by how long the system can follow the motion determined by the Schrödinger equation without being disturbed, and it is an important index in designing quantum circuits.

**Keywords: nuclear magnetic resonance (NMR) computer; ion trap computer; quantum dot computer.**

## 11.1 Nuclear magnetic resonance (NMR) computers

### 11.1.1 *Principles of NMR computers*

The **nuclear magnetic resonance computer** (the **NMR computer**) makes use of nuclear magnetic resonance (NMR); nuclear spins in molecules are adopted as quantum bits and controlled by external magnetic fields. As seen in Fig. 11.1, when an atomic nucleus is placed in a strong static

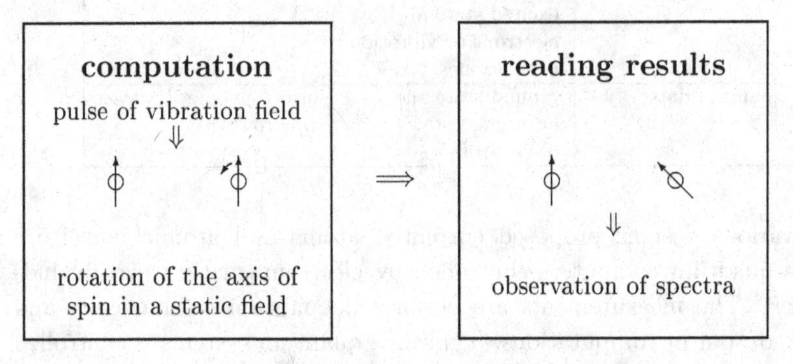


Fig. 11.1 NMR computer. "Computation" by applying vibrating magnetic field (microwave) to nuclear spins. When the computation is finished, the "results" are extracted by the measurement of absorption and emission of microwave from nuclei.

magnetic field, the energy levels are split according to the directions of the spins. This phenomenon is called the **Zeeman effect**. If one applies comparatively weak vibrating magnetic field perpendicular to the static field, the spin axis rotates if the resonant frequency coincides with the vibration frequency. By giving this vibrating field in pulses, like

it is possible to select the spin to rotate, adjusting the frequency and the period. For example, one can make a mixed spin state by rotating the spin angle 90° or make the negation gate rotating the spin by 180°. The period of vibration of magnetic field is of the order of ten nanoseconds ($\sim 10^{-8}$ s), while the duration of application of field is of the order of a hundred microseconds ($\sim 10^{-4}$ s) in experiments. While the decoherence time depends on the condition, it is considered to be the order of seconds ($\sim 1$ s). Therefore, it is possible to perform a large number of logical operations on the quantum circuits before decoherence. For example, in

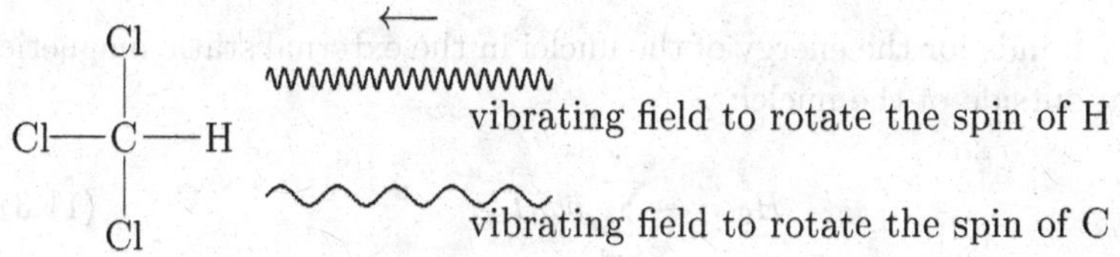

Cl
Cl—C—H
Cl
vibrating field to rotate the spin of H
vibrating field to rotate the spin of C

Fig. 11.2 Mechanism of NMR with chloroform. The nuclei of hydrogen $^{1}$H and carbon $^{13}$C, both have spin 1/2, are used as qubits. These nuclei are controlled separately by vibrating magnetic fields of different frequency.

chloroform shown in Fig. 11.2, the nucleus of hydrogen $^{1}$H and the nucleus of carbon $^{13}$C have both spin 1/2, and they act as quantum bits.

The values $|0\rangle$ and $|1\rangle$ of a qubit can be associated to the upward and downward directions of the nuclear spin, so that nuclei with spin magnitude 1/2 are suitable to this end. For example, nuclei such as $^{1}$H, $^{13}$C, $^{14}$N, $^{19}$F, $^{31}$P, which have spin 1/2 in the ground state, have been tested in experiments. If a molecule has two nuclei, a quantum computer of two bits can be implemented. If it has four nuclei, a computer of four bits can be realized. In real experiments, it is hard to test on separate molecules or nuclei. Instead, ensembles of them, for example, molecules in liquid, are used. Then, statistical-mechanical treatment is needed for the order of Avogadro number of molecules in room temperature . In this respect, the situation is much different from that for ion traps described in Section 11.2.

### 11.1.2 *NMR and spin rotations*

It is known experimentally that for a nucleus with the spin $I$ and its $z$-component $I_z$, its states are split into the Zeeman levels with the energy

$$E = -gB_0 I_z \tag{11.1}$$

when the static magnetic field $B_0$ is applied. This energy splitting (11.1) distinguished by the spin direction is called the **hyperfine structure**. In Eq. (11.1), the coefficient $g$ is called the **gyromagnetic ratio** ($g$-factor) which is the constant determined by the ratio of magnetic moment $\mu$ to the spin of the nucleus. In a molecule, due to the shielding effect by electrons, the $g$-factor has an effective value different from that in the isolated nucleus.

Under the static magnetic field of the direction to the $z$-axis, the Hamiltonian of the system of $N$ nuclei is given as

$$H = H_0 + H_1, \tag{11.2}$$

where $H_0$ stands for the energy of the nuclei in the external static magnetic field from outside of the nuclei:

$$H_0 = -\sum_{i=1}^{N} \hbar\omega_i I_{zi}, \tag{11.3}$$

where $\hbar\omega_i \equiv \gamma_i B_0$ gives the energy difference between the spin-up and spin-down states. The factor $\gamma_i$ means the $g$-factor of the $i$th nucleus. Each nucleus has the different value $\gamma_i$ depending on the configuration of electrons around it. The second term, $H_1$, is the interaction between nuclei with spins $I_{zi}$ and $I_{zj}$ and is approximated as

$$H_1 = 2\pi\hbar \sum_{i<j} J_{ij} I_{zi} I_{zj}, \tag{11.4}$$

where $J_{ij}$ is the strength of the interaction. The energies involved in the hyperfine structure correspond to those of radiowave of several hundred Hz, which is easy to control.

Let us think about specific methods to control quantum bits by changing the direction of the spin. In order to control the direction of the spin, one applies the alternating magnetic field of several hundred Hz along the direction in the $xy$-plane (the spin direction is along the $z-$axis. This operation introduces an additional term of oscillating magnetic field to the Hamiltonian of Eq. (11.2). To see how this oscillating magnetic field works, let us take a nucleus with spin 1/2. For simplicity, we may ignore $H_1$. If the oscillating magnetic field is rotating in the $xy$-plane, the Hamiltonian is written as

$$H = -\gamma\left(B_0 s_z + B_1((\cos\omega t)s_x - (\sin\omega t)s_y)\right), \tag{11.5}$$

where $s_x$, $s_y$,and $s_z$ are three components of the spin operator. The Schrödinger equation is given as

$$i\hbar\frac{d}{dt}|\psi(t)\rangle = H|\psi(t)\rangle. \tag{11.6}$$

Since $s_x$, $s_y$ and $s_z$ are $2\times 2$ matrices, the state $|\psi(t)\rangle$ is also given by the two-dimensional vector. To solve (11.6), we introduce a state $|\phi(t)\rangle$ by

$$|\psi(t)\rangle \equiv e^{i\omega t s_z}|\phi(t)\rangle, \tag{11.7}$$

and solve the equation with respect to it. The left-side hand of Eq. (11.6) is given by

$$i\hbar e^{i\omega t s_z}\left(i\omega s_z|\phi(t)\rangle + \frac{d}{dt}|\phi(t)\rangle\right), \tag{11.8}$$

while the right-hand side becomes

$$(-\gamma B_0 e^{i\omega t s_z} s_z - \gamma B_1\cos(\omega t)s_x e^{i\omega t s_z} + \gamma B_1 \sin(\omega t) s_y e^{i\omega t s_z})|\phi(t)\rangle. \tag{11.9}$$

Multiplying both sides by $e^{-i\omega t s_z}$ from the left, and using the following formulas (See Example 11.1):

$$e^{-i\omega t s_z} s_x e^{i\omega t s_z} = \cos(\omega t)s_x + \sin(\omega t)s_y, \tag{11.10}$$

$$e^{-i\omega t s_z} s_y e^{i\omega t s_z} = \cos(\omega t)s_y - \sin(\omega t)s_x, \tag{11.11}$$

we obtain

$$i\hbar\frac{d}{dt}|\phi(t)\rangle = -\left[(\gamma B_0 - \hbar\omega)s_z + \gamma B_1 s_x\right]|\phi(t)\rangle. \tag{11.12}$$

By integrating Eq. (11.12), we have

$$|\phi(t)\rangle = \exp\left\{\frac{it}{\hbar}\left[(\gamma B_0 - \hbar\omega)s_z + \gamma B_1 s_x\right]\right\}|\phi(0)\rangle. \tag{11.13}$$

In particular, if the following condition holds between $\omega$ of the oscillating field and $B_0$ of the static field,

$$\hbar\omega = \gamma B_0, \tag{11.14}$$

the state vector $|\phi(t)\rangle$ rotates about the $x$-axis with angular frequency $\gamma B_1/\hbar$.

$$|\phi(t)\rangle = \exp\left(\frac{i\gamma B_1 t}{\hbar}s_x\right)|\phi(0)\rangle. \tag{11.15}$$

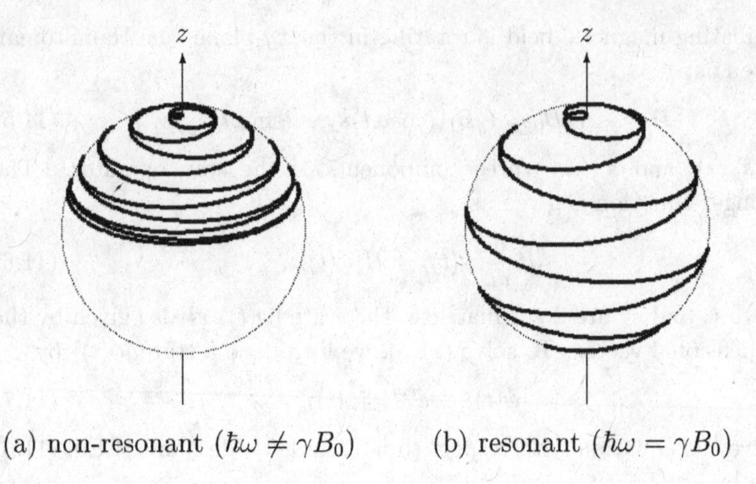


Fig. 11.3 Change of the direction of spin by the oscillating magnetic field. The full lines show how the direction of spin changes with time. The spin is placed along the positive $z$-axis at $t = 0$. Usually, $B_0$ is several hundred MHz, while $B_1$ may take several ten kHz to have the resonance. Since $B_0 \gg B_1$, the speed of rotation about the $z$-axis is higher than that of the change of the polar angle. The case (b) shows the resonance where the spin can be changed to any direction.

Using Eq. (11.15), we obtain the wave function of the nuclear spin (11.7) as

$$|\psi(t)\rangle = \exp\left(\frac{i\gamma B_0 t}{\hbar} s_z\right) \exp\left(\frac{i\gamma B_1 t}{\hbar} s_x\right) |\psi(0)\rangle, \tag{11.16}$$

which is interpreted as the combination of the rotational motion around the $z$-axis and the rotation around the $x$-axis. Since the rotational axis of spinning object itself rotates, this motion is similar to the precession of a top. Incidentally, Eq. (11.14) corresponds to the resonance between the oscillating field and the nuclear spin in the static field, where the energy absorption from the oscillating field becomes maximum. Figure 11.3(b) shows the case where the spin can be in any direction.

---

**Example 11.1** For the spin 1/2 operators, derive the following formulas that are used in p. 185:

$$e^{-i\lambda s_z} s_x e^{i\lambda s_z} = (\cos\lambda)\, s_x + (\sin\lambda)\, s_y, (11.10) \tag{11.10}$$

$$e^{-i\lambda s_z} s_y e^{i\lambda s_z} = -(\sin\lambda)\, s_x + (\cos\lambda)\, s_y (11.11) \tag{11.11}$$

where $\lambda$ is an arbitrary number.

**Solution** Equation (11.10) is obtained by setting $\alpha = -\lambda$ in Eq.

(3.81). Equation (11.11) is also obtained in a similar manner. As another derivation, let us consider the left-hand side as a function $F(\lambda)$ of $\lambda$, and differentiate it:

$$\begin{aligned} F'(\lambda) &= -e^{-i\lambda s_z}\, is_z s_x\, e^{i\lambda s_z} + e^{-i\lambda s_z}\, s_x i s_z\, e^{i\lambda s_z} \\ &= -ie^{i\lambda s_z}\, \overbrace{[s_z, s_x]}^{is_y}\, e^{-i\lambda s_z} \\ &= e^{i\lambda s_z}\, s_y\, e^{-i\lambda s_z}. \end{aligned} \tag{11.17}$$

By differentiating it once more, we obtain

$$\begin{aligned} F''(\lambda) &= -e^{-i\lambda s_z}\, is_z s_y\, e^{i\lambda s_z} + e^{-i\lambda s_z}\, s_x i s_z\, e^{i\lambda s_z} \\ &= -ie^{-i\lambda s_z}\, \overbrace{[s_z, s_y]}^{-is_x}\, e^{i\lambda s_z} \\ &= -e^{-i\lambda s_z}\, s_x\, e^{i\lambda s_z} = -F(\lambda). \end{aligned} \tag{11.18}$$

Equation (11.18) is similar to the equation of harmonic oscillator, for which the general solution is written as

$$F(\lambda) = (\cos\lambda)A + (\sin\lambda)B, \tag{11.19}$$

where $A$ and $B$ are $2 \times 2$ matrices of constants independent of $\lambda$. The matrix $A = s_x$ is obtained by substituting 0 for $\lambda$ in Eqs. (11.10) and (11.19), while $B = s_y$ is obtained by substituting 0 for $\lambda$ in both Eq. (11.17) and the first derivative of Eq. (11.19) with respect to $\lambda$. Equation (11.11) can be derived in a similar manner.

---

**Example 11.2** Static magnetic field of 10 T (Tesla) is applied to a hydrogen nucleus. Estimate the frequency of the electromagnetic wave that resonates with the two levels with the hyperfine structure. The magnetic dipole moments of hydrogen $^1$H and carbon $^{13}$C are $\mu(^1\mathrm{H}) \approx 2.8\ \mu_\mathrm{N}$ and $\mu(^{13}\mathrm{C}) \approx 0.7\ \mu_\mathrm{N}$ respectively, where $\mu_\mathrm{N}$ is the magnetic dipole moment of a proton, $\mu_\mathrm{N} = e\hbar/(2m_\mathrm{p})$. The symbol $m_\mathrm{p}$ stands for the mass of a proton. The value of nuclear magneton is $5.05 \times 10^{-27}$ JT$^{-1}$. The Planck constant is given by $h = 6.63 \times 10^{-34}$ Js. Consider also the case of $^{13}$C ($g = 1.4\ \mu_\mathrm{N}$).

**Solution** As the $g$-factor of hydrogen, we take the value $g = \mu/s = 5.6\ \mu_\mathrm{N}$. From Eq. (11.14), the frequency $h\nu = gB_0$ becomes

$$\nu = \frac{gB_0}{h}. \tag{11.20}$$

In the case of $^1$H, we have

$$\nu = \frac{5.6 \times 5.05 \times 10^{-27}\mathrm{JT}^{-1} \times 10\mathrm{T}}{6.63 \times 10^{-34}\mathrm{Js}} = 427 \times 10^6 \mathrm{s}^{-1}, \tag{11.21}$$

namely, about 430 MHz. In the case of $^{13}$C, we have

$$\nu = \frac{1.4 \times 5.05 \times 10^{-27}\mathrm{JT}^{-1} \times 10\mathrm{T}}{6.63 \times 10^{-34}\mathrm{Js}} = 107 \times 10^{6}\mathrm{s}^{-1}, \tag{11.22}$$

namely, about 110 MHz.

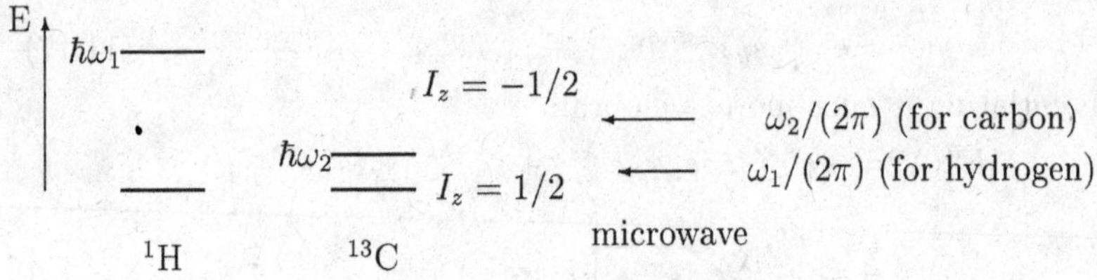

---

E

$\hbar\omega_1$

$\hbar\omega_2$

$I_z = -1/2$

$I_z = 1/2$

$\omega_2/(2\pi)$ (for carbon)

$\omega_1/(2\pi)$ (for hydrogen)

microwave

$^{1}$H $^{13}$C

Fig. 11.4 The energies $H_0$ of the hyperfine structure in hydrogen and carbon. The microwaves resonating with them are given by Eqs. (11.21) and (11.22) for $^1$H and $^{13}$C, respectively.

If one applies the alternating magnetic field of frequency corresponding to the energy difference between the spin-up and spin-down states, the direction of the spin axis will change depending on the time of application. Figure 11.4 shows the energy levels of spins up and down. One can control each spin separately by applying the microwave of appropriate frequency $\omega_i$. When one assigns the logical values "0" and "1" to the two states in Fig. 11.4, one can observe the logical value 0 or 1 by measuring the absorption and emission spectra when the computation finishes, as shown in Fig. 11.5.

### 11.1.3 *Statistical treatment*

The NMR computer is implemented by the chloroform solution as Fig. 11.6 and can be controlled by applying the external magnetic field. We must take into account that Avogadro's number of molecules are involved as well as the effect of heat since experiments are carried out at room temperature. According to statistical mechanics for the system in thermal equilibrium with absolute temperature $T$, the probability of the state with energy $E_i$ is proportional to the factor $\exp(-E_i/(k_\mathrm{B}T))$, where $k_\mathrm{B} = 1.381\times10^{-23}$ JK$^{-1}$ is the Boltzmann constant. Therefore the expectation value of a physical observable $O$ becomes

$$\langle O\rangle = \frac{\sum_i O_i \exp(-E_i/(k_\mathrm{B}T))}{\sum_i \exp(-E_i/(k_\mathrm{B}T))}. \tag{11.23}$$

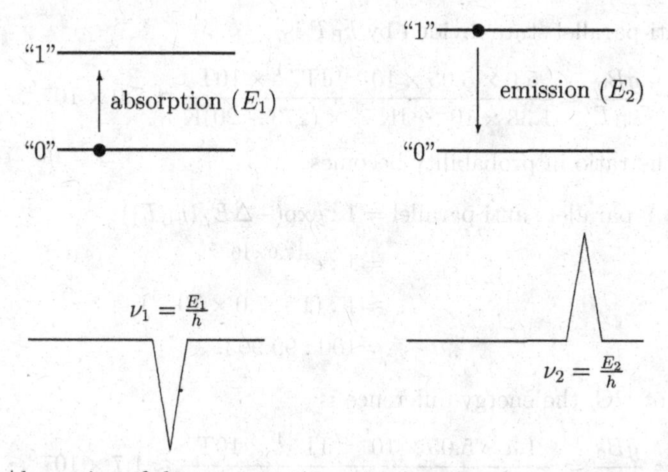


Fig. 11.5 Absorption and emission of electromagnetic wave in spectra and their relations with logical values.

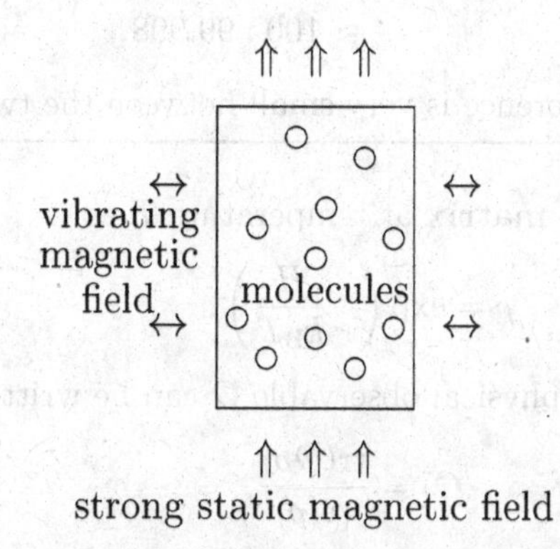


Fig. 11.6 Nuclear magnetic resonance (NMR) computer. chloroform solution is used as qubits.

---

**Example 11.3** Nuclei $^1$H and $^{13}$C are placed in a static magnetic field of flux $B_0 = 10$ T. At temperature 20°, calculate the ratios of spin parallel to anti-parallel states to the magnetic field.

**Solution** In $^1$H, the energy difference $\Delta E$ between the spin parallel and spin anti-parallel state divided by $k_{\mathrm{B}}T$ is

$$\frac{\Delta E}{k_{\mathrm{B}}T} = \frac{gB_0}{k_{\mathrm{B}}T} = \frac{5.6 \times 5.05 \times 10^{-27}\mathrm{JT}^{-1} \times 10\mathrm{T}}{1.38 \times 10^{-23}\mathrm{JK}^{-1} \times (273+20)\mathrm{K}} \approx 7.0 \times 10^{-5}.$$

Therefore, the ratio in probability becomes

$$\begin{aligned}
\text{parallel : anti parallel} &= 1 : \exp(-\Delta E/(k_{\mathrm{B}}T)) \\
&= 1 : e^{-7.0\times 10^{-5}} \\
&\approx 1 : (1 - 7.0 \times 10^{-5}) \\
&= 100 : 99.993.
\end{aligned}$$

In the case of $^{13}$C, the energy difference is

$$\frac{\Delta E}{k_{\mathrm{B}}T} = \frac{gB_0}{k_{\mathrm{B}}T} = \frac{1.4 \times 5.05 \times 10^{-27}\mathrm{JT}^{-1} \times 10\mathrm{T}}{1.38 \times 10^{-23}\mathrm{JK}^{-1} \times (273+20)\mathrm{K}} \approx 1.7 \times 10^{-5},$$

and the ratio becomes

$$\begin{aligned}
\text{parallel : anti parallel} &\approx 1 : e^{-1.7\times 10^{-5}} \\
&\approx 1 : (1 - 1.7 \times 10^{-5}) \\
&\approx 100 : 99.998.
\end{aligned}$$

It turns out that the difference is very small between the two cases.

---

Defining the **density matrix** at temperature $T$ as

$$\rho = \exp\left(-\frac{H}{k_{\mathrm{B}}T}\right), \tag{11.24}$$

the expectation value of physical observable $O$ can be written as

$$\langle O \rangle = \frac{\mathrm{tr}(O\rho)}{\mathrm{tr}(\rho)}. \tag{11.25}$$

At room temperature, because $\langle H \rangle / k_{\mathrm{B}}T \sim 10^{-4}$, the density matrix of the system of $n$ articles with spin 1/2 can be approximated with good accuracy as

$$\rho \approx 2^{-n}[1 - H/k_{\mathrm{B}}T]. \tag{11.26}$$

The factor $2^{-n}$ reflects the degree of freedom of the system. For the system with a single spin 1/2 state ($n = 1$), we have

$$\rho = \frac{1}{2} + \frac{\hbar\omega_1}{4k_{\mathrm{B}}T}\begin{bmatrix} 1 & 0 \\ 0 & -1 \end{bmatrix}. \tag{11.27}$$

if the energy difference between the ground state and an excited state is $\hbar\omega_1$. In a system with two spin 1/2 states having the degree of freedom $2^2$, we have

$$\rho = \frac{1}{4} + \frac{1}{8k_\mathrm{B}T}\begin{bmatrix} \hbar\omega_1 + \hbar\omega_2 & 0 & 0 & 0 \\ 0 & -\hbar\omega_1 + \hbar\omega_2 & 0 & 0 \\ 0 & 0 & +\hbar\omega_1 - \hbar\omega_2 & 0 \\ 0 & 0 & 0 & -\hbar\omega_1 - \hbar\omega_2 \end{bmatrix}. \tag{11.28}$$

if the energy differences are $\hbar\omega_1$ and $\hbar\omega_2$ for the first and second spins, respectively, as shown in Fig. 11.4.

First, the initial state must be set at the beginning of computation. Though it is natural to start from the ground states, it is not easy to bring all spins to the ground state simultaneously, because $\Delta E \ll k_\mathrm{B}T$ for typical energy differences $\Delta E$ and the room temperature $T$. Instead, an ingenious way was devised to extract the results from the initial ground state only even though other excited states are mixed. In the time-average method shown in Fig. 11.8, one varies the distribution of excited states, and by taking the time average of the experimental results, it is possible to kill the effect from the excited states on the final results.

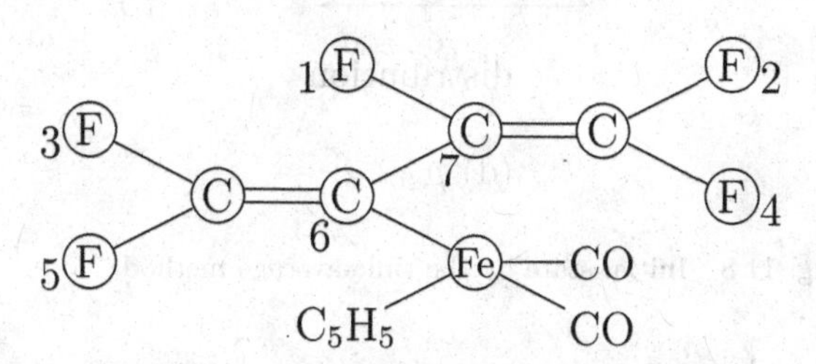


Fig. 11.7 Molecule used for the factorization of an integer 15.

### 11.1.4 *Experiment of quantum factorization*

In December of 2001, the successful experiment was reported on the factorization of an integer 15 using a seven-bit NMR computer of the molecule shown in Fig. 11.7.[1] This experiment was performed by the method illustrated in Example 8.2 in Chapter 8, where five $^{19}$F and two $^{13}$C were used

[1] Lieven M. K. Vandersypen, Matthias Steffen, Gregory Breyta, Costatino S. Yannoni, Mark H. Sherwood and Isaac L. Chuang, Experimental realization of Shor's quantum factoring algorithm using nuclear magnetic resonance, Nature **414**, 883 (Dec. 2001).

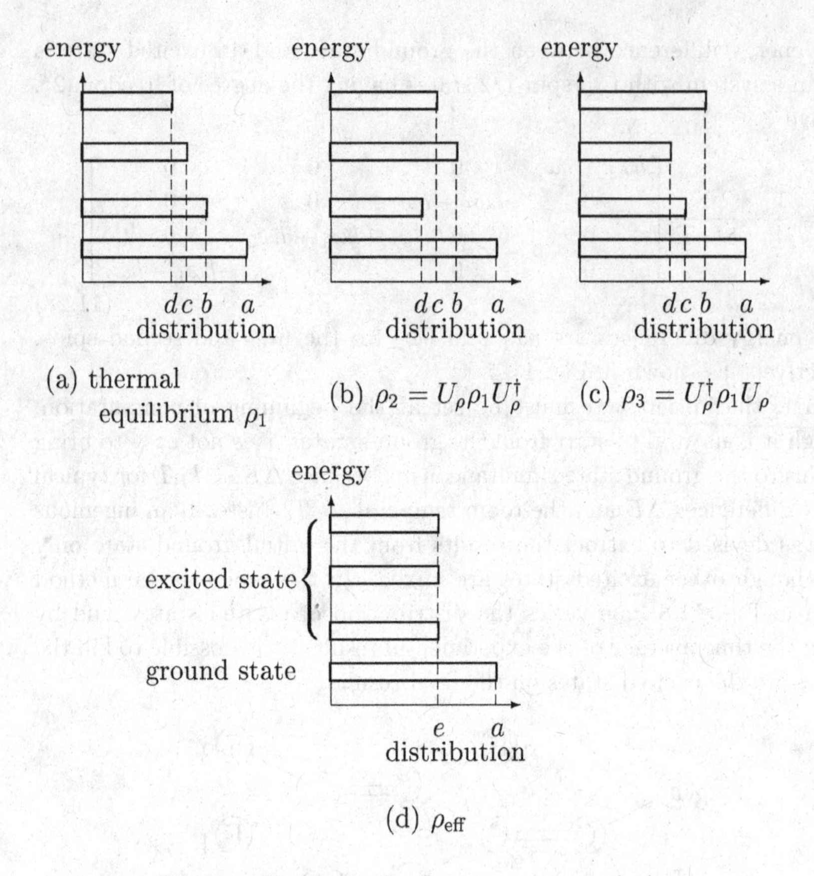


Fig. 11.8 Initial state by the time-average method.

as the qubits. Although several nuclei have the same atomic numbers, each nucleus independently can act as a different qubit because it has the different effective $g$-factor depending on the configuration of the atoms around it.

Since the experiment is performed at room temperature, excited states are largely mixed in the initial states. In order to remove the effect from excited states, the time-average method is adopted as shown in Fig. 11.8. Then, one can practically start the computation from the ground state. Let us consider how the time-average method works in the system of two spin states. Taking account of the form of Eq. (11.28), let us write the density matrix at room temperature as

$$\rho_1 = \begin{pmatrix} a & 0 & 0 & 0 \\ 0 & b & 0 & 0 \\ 0 & 0 & c & 0 \\ 0 & 0 & 0 & d \end{pmatrix}, \tag{11.29}$$

where $a + b + c + d = 1$. We also define a unitary transformation

$$U_\rho = \begin{pmatrix} 1 & 0 & 0 & 0 \\ 0 & 0 & 1 & 0 \\ 0 & 1 & 0 & 0 \\ 0 & 0 & 0 & 1 \end{pmatrix} \begin{pmatrix} 1 & 0 & 0 & 0 \\ 0 & 1 & 0 & 0 \\ 0 & 0 & 0 & 1 \\ 0 & 0 & 1 & 0 \end{pmatrix} \tag{11.30}$$

which can be obtained by combining two controlled-NOT gates such as Eq. (6.25). We apply $U_\rho$ and $U_\rho^\dagger$ on both sides of $\rho_1$,

$$\rho_2 = U_\rho \rho_1 U_\rho^\dagger = \begin{pmatrix} a & 0 & 0 & 0 \\ 0 & d & 0 & 0 \\ 0 & 0 & b & 0 \\ 0 & 0 & 0 & c \end{pmatrix}. \tag{11.31}$$

Then, by interchanging $U_\rho$ and $U_\rho^\dagger$, we have

$$\rho_3 = U_\rho^\dagger \rho_1 U_\rho = \begin{pmatrix} a & 0 & 0 & 0 \\ 0 & c & 0 & 0 \\ 0 & 0 & d & 0 \\ 0 & 0 & 0 & b \end{pmatrix}. \tag{11.32}$$

In the time-average method, one performs the experiments three times corresponding to the density matrices Eqs. (11.29), (11.31) and (11.32), and then averages the results of measurements. The averaged density matrix $\rho_{\text{eff}}$ is expressed as

$$\begin{aligned} \rho_{\text{eff}} &= \frac{1}{3}(\rho_1 + \rho_2 + \rho_3) = \begin{pmatrix} a & 0 & 0 & 0 \\ 0 & e & 0 & 0 \\ 0 & 0 & e & 0 \\ 0 & 0 & 0 & e \end{pmatrix} \\ &= e \begin{pmatrix} 1 & 0 & 0 & 0 \\ 0 & 1 & 0 & 0 \\ 0 & 0 & 1 & 0 \\ 0 & 0 & 0 & 1 \end{pmatrix} + (a - e) \begin{pmatrix} 1 & 0 & 0 & 0 \\ 0 & 0 & 0 & 0 \\ 0 & 0 & 0 & 0 \\ 0 & 0 & 0 & 0 \end{pmatrix}, \end{aligned} \tag{11.33}$$

where $e \equiv (b+c+d)/3$. The effects from the excited states can be cancelled out as the overall uniform background, and thus the experimental result with the ground state as the initial state is obtained with probability $(a-e)$.

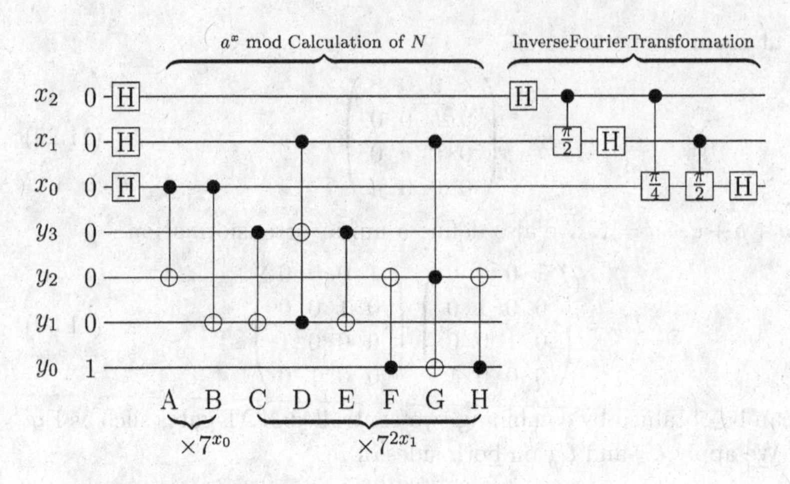


Fig. 11.9 Quantum circuit of Shor's algorithm. The three qubits $x_0$ to $x_2$ constitute the first register, while the four qubits $y_0$ to $y_3$ constitute the second register. After the inverse Fourier transformation, the first register is measured.

The quantum circuit for the factorization of the integer 15 was designed by a seven-bit NMR computer as shown in Fig. 11.9. The three qubits $x_0$ to $x_2$ constitute the first register, while the four qubits $y_0$ to $y_3$ constitute the second register. First, one sets $x = 0$ and $y = 1$. Then, the Hadamard transformations are applied to the first register. The gates A to H perform the congruence operation $a^x$ (mod $N$) in the following way. The value $a^x$ can be decomposed as

$$\begin{aligned} a^x &= a^{x_0+2x_1+4x_2} \\ &= a^{x_0} \times (a^2)^{x_1} \times (a^4)^{x_2}. \end{aligned} \tag{11.34}$$

By substituting the adopted values $a = 7$ and $N = 15$, the factors can be reduced as

$$7^2 = 49 \equiv 4 \pmod{15}, \tag{11.35}$$

$$7^4 = (7^2)^2 \equiv 4^2 = 16 \equiv 1 \pmod{15}, \tag{11.36}$$

namely, the congruence is converted to be

$$a^x \equiv 7^{x_0} \times 4^{x_1} \pmod{15}. \tag{11.37}$$

As a result of two controlled-NOT gates A and B, the value of $7^{x_0}$ is put in $y$. This is because the set of gates A and B gives the result:

- if $x_0 = 0$, $y$ keeps 1 unchanged;
- if $x_0 = 1$, $y$ is 7.

In the next steps, C to H, $y$ is multiplied by $7^{2x_1} \equiv 4^{x_1}$. To understand it, we note

$$\begin{aligned} 4y &= 4(y_0 + 2y_1 + 4y_2 + 8y_3) \\ &= 4y_0 + 8y_1 + 16y_2 + 32y_3 \\ &= 4y_0 + 8y_1 + (15+1)y_2 + (2 \times 15 + 2)y_3 \\ &\equiv y_2 + 2y_3 + 4y_0 + 8y_1 \pmod{15}. \end{aligned} \tag{11.38}$$

That is to say, multiplying a four-bit number $y$ by 4 modulo 15 is equivalent to the re-arrangement of bits: $y_3y_2y_1y_0 \to y_1y_0y_3y_2$. Therefore, multiplying $y$ by $4^{x_1}$ is attained by the following operations:

- if $x_1 = 0$, $y$ is unchanged;
- if $x_1 = 1$, interchange $y_1$ and $y_3$, and also interchange $y_0$ and $y_2$.

The three gates from C to E perform the controlled interchange of $y_1$ and $y_3$, with $x_1$ acting as the control bit, as explained as follows.

- if $x_1 = 0$, D makes nothing. Thus it is clear that $y_3$ is unchanged. Though $y_1$ is controlled by $y_3$ through C and E, its value is not changed because
$$y_1 \overset{\text{C}}{\to} y_3 \oplus y_1 \overset{\text{E}}{\to} y_3 \oplus (y_3 \oplus y_1) = (y_3 \oplus y_3) \oplus y_1 = 0 \oplus y_1 = y_1.$$
- If $x_1 = 1$, D acts as the controlled-NOT gate with $y_1$ as the control bit, and $y_3$ as the target bit. Therefore, by the series of gates, $y$ follows the process:
$$\begin{matrix} y_3 \\ y_1 \end{matrix} \overset{\text{C}}{\Rightarrow} \begin{matrix} y_3 \\ y_3 \oplus y_1 \end{matrix} \overset{\text{D}}{\Rightarrow} \begin{matrix} y_3 \oplus (y_3 \oplus y_1) = y_1 \\ y_3 \oplus y_1 \end{matrix} \overset{\text{E}}{\Rightarrow} \begin{matrix} y_1 \\ y_1 \oplus (y_3 \oplus y_1) = y_3. \end{matrix}$$
As a result, $y_3$ and $y_1$ are interchanged.

In other words, the set of gates C$\to$E acts as the Fredkin gate (see Section 6.2) with the control bit $x_1$. In a similar way, we see that the gates F$\to$H act as the Fredkin gate interchanging $y_2$ and $y_0$ with the control bit $x_1$. In this way, we have seen that the set of gates C$\to$H performs the multiplication by $(7^2)^{x_1}$ modulo 15. After this operation, one applies the inverse Fourier transformation to the first register, and then one measures the result to obtain the factor.

The decoherence may happen by the influence from outside of the system, and also by the interactions between molecules. The decoherence time is typically 0.5 to 10 seconds in the system of Fig. 11.9. In the experiments, it was possible to finish the quantum computation in less time than the decoherence.

## 11.2 Ion trap computers

### 11.2.1 *Basic principles*

**The ion trap** is a technique to confine ionized atoms in a small space by electromagnetic forces. To apply this technique to computers, one controls each atom by laser as shown in Fig. 11.10. As qubits one uses the eigenstates

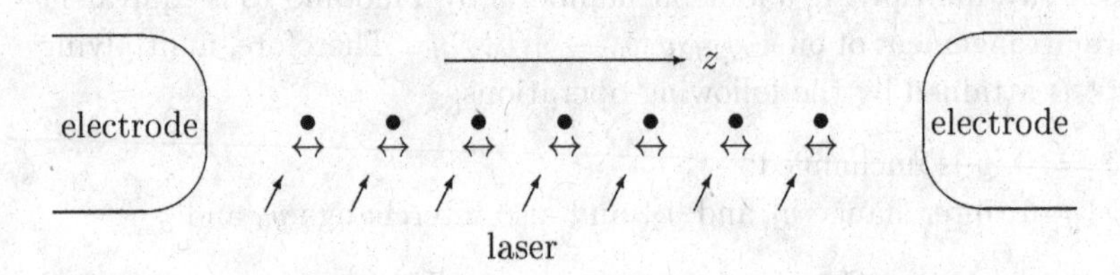


Fig. 11.10 Linear ion trap. The symbol • shows atoms (ions) placed in a line. The laser controls the state of each atom.

of electrons in each atom. The quantum states of vibration of the atoms can also be utilized as qubits. Figure 11.11 illustrates the procedure of computation using this technique.

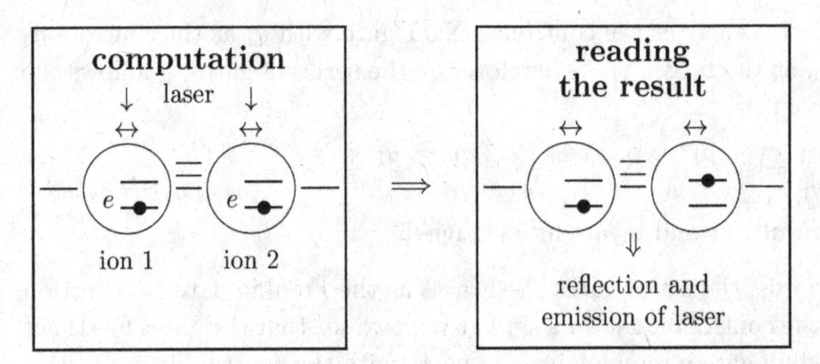


Fig. 11.11 Ion trap computer. The logical values are represented by the electronic states in the ionized atoms. Computation is controlled by laser. When computation is finished, one measures the light absorbed by or emitted from atoms to detect the result.

### 11.2.2 *Ion trap*

An atom becomes an ion when it loses or gets electrons. The ion is a particle charged positive or negative. Since the ion is not neutral, it is possible to accelerate it or bend its orbit by applying electromagnetic field from outside. However, it is hard to stop it at a fixed point. According to Maxwell's equations, the Laplacian of the electric field vanishes in vacuum without any charges so that the potential for a charged particle can not be the minimum. Therefore it is not possible to stop the ion by the electrostatic field in vacuum. The ion trap is the technique to stop the ion by neat methods. Among several methods, so called, the Paul trap method uses the technique of changing the direction of alternating eclectic field rapidly to capture the ion.

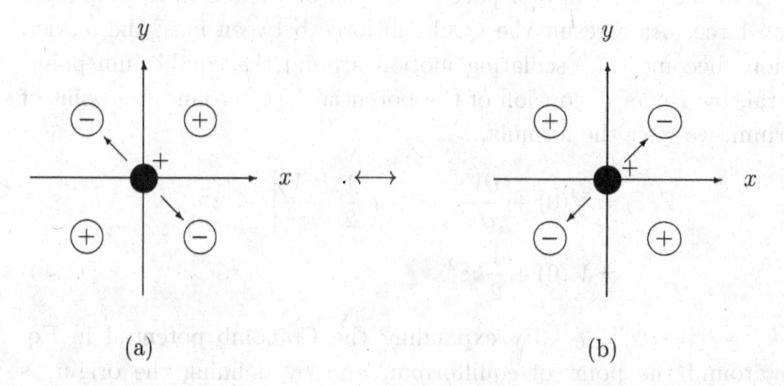


Fig. 11.12 Ion trap by the alternating eclectic field. The filled circles at the center represent the ions, while the symbols $\oplus$ and $\ominus$ around them show positively and negatively charged electrodes, respectively. The electrodes change the signs quickly all the time. In case (a), the positive ion try to go toward the pole $\ominus$, but it can not continue to move to any fixed direction because the the electrode quickly changes the sign into $\oplus$. The ion trap is the method to confine the ion in a small region around a fixed point by rapidly changing the sign of electric poles.

As shown in Fig. 11.12, a positive ion, which tends to go in the direction of lower electric potential, can not move enough to any fixed direction because the potential is inversed rapidly. Then, one can confine the ion in a small space. As an effective potential to represent the rapidly alternating ion-confining electric potential, one may use the harmonic oscillator potential in the $xy$-plane. For the direction of the $z$-axis, one can limit the moving space of the ion putting two electrodes with the same charges as that of the ion both up and down of the ions. Arranging that the characteristic frequency in the $xy$ plane is much larger than that along the $z$-axis, we ignore the motion parallel to the $xy$-plane hereafter.

In the system where $N$ ions are arranged along the $z$-axis as shown in

Fig. 11.10, the Hamiltonian can be written as

$$H = \sum_{i}^{N} \left( \frac{1}{2m} p_i^2 + \frac{1}{2} k' z_i^2 + H_{\text{in}} \right) + \sum_{j>i}^{N} \frac{e^2}{4\pi\epsilon_0 |z_j - z_i|}, \tag{11.39}$$

where $m$ is the mass of each ion while $k'$ is the spring constant with respect to the motion along the $z$-axis. The intrinsic motion within each ion is represented by $H_{\text{in}}$, while the last summation stands for the Coulomb potential between the ions.

Two ions have tendency to part from each other due to the repulsive Coulomb force. As a result the Coulomb force between ions, the motion of the ions becomes an oscillating motion around the equilibrium point. In general, by Taylor expansion of the potential $V(z)$ around the point of equilibrium, we have the formula,

$$\begin{aligned} V(z) &\approx V(0) + \left. \frac{\partial V}{\partial z} \right|_{z=0} z + \frac{1}{2} \left. \frac{\partial^2 V}{\partial z^2} \right|_{z=0} z^2 \\ &= V(0) + \frac{1}{2} k z^2, \end{aligned} \tag{11.40}$$

where $k = \partial^2 V / \partial z^2 |_{z=0}$. By expanding the Coulomb potential in Eq. (11.39) around the point of equilibrium, and by defining the origin as $V(0) = 0$, we have a linearly approximated Hamiltonian:

$$H = \sum_{i=1}^{N} \left( \frac{p_i^2}{2m} + \frac{1}{2} k' z_i^2 \right) + \sum_{j>i}^{N} \frac{1}{2} k (z_j - z_i)^2. \tag{11.41}$$

If the coordinates are arranged in accordance with the normal modes, the system is converted to be a set of harmonic oscillators. Thus, the energies of oscillation of the ions are classified to be the center of mass motion, the **breathing mode** and the harmonic oscillator mode as shown in Fig. 11.13(b). For the quantum computer, one can use the ground state and the first-excited state in each ion, as well as the ground state and low-lying states of the oscillation of the ion system.

---

**Example 11.4** As shown in Fig. 11.14, two particles of mass $m$ connected by springs are allowed to move in one dimensional space. Calculate the characteristic frequencies of the system.

**Solution** We will derive the classical equations of motion as follows. Let us write the displacement from the equilibrium points of the two particles as $z_1(t)$ and $z_2(t)$ as a function of time $t$. The spring constant between

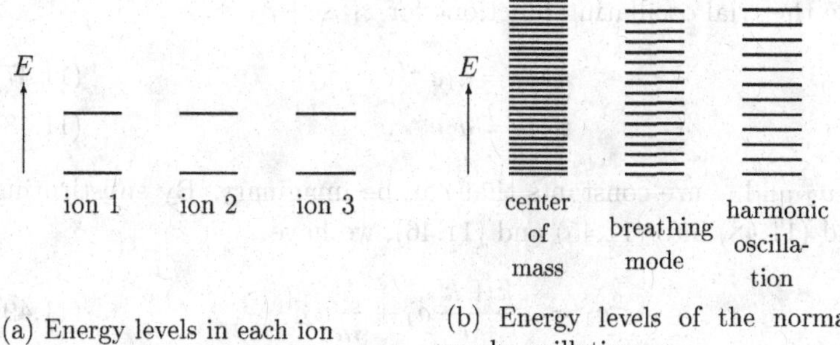


Fig. 11.13 Energy levels in the system of ion traps. (a) shows the electric energy levels in each ion while (b) shows the levels arising from the oscillations among ions. The sum of all these modes constitutes the energy of the total system.

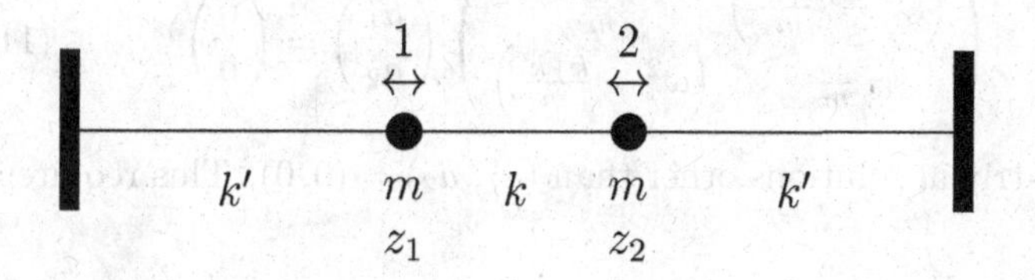


Fig. 11.14 Oscillation of a system of two particles.

two particles is denoted by $k$, while that between the wall and the particle is given by $k'$. The potential energy of the system can be written as

$$V(z_1, z_2) = \frac{1}{2}k'z_1^2 + \frac{1}{2}k(z_1 - z_2)^2 + \frac{1}{2}k'z_2^2. \qquad (11.42)$$

Since the forces acting upon the two particles are derived as $F_1 = -dV/dz_1$ and $F_2 = -dV/dz_2$ from the potential energy of the springs, the Newton's equations of motion become

$$m\frac{d^2 z_1}{dt^2} = -k'z_1 + k(z_2 - z_1), \qquad (11.43)$$

$$m\frac{d^2 z_2}{dt^2} = k(-z_2 + z_1) - k'z_2, \qquad (11.44)$$

from which we have coupled linear differential equations for $z_1$ and $z_2$:

$$\frac{d^2 z_1}{dt^2} = -\frac{k + k'}{m}z_1 + \frac{k}{m}z_2, \qquad (11.45)$$

$$\frac{d^2 z_2}{dt^2} = \frac{k}{m}z_1 - \frac{k + k'}{m}z_2. \qquad (11.46)$$

Let us take the trial oscillating functions for $z_1$ and $z_2$,

$$z_1(t) = a_1 e^{i\omega t}, \tag{11.47}$$

$$z_2(t) = a_2 e^{i\omega t}, \tag{11.48}$$

where $a_1$, $a_2$ and $\omega$ are constants that can be imaginary. By substituting (11.47) and (11.48) into (11.45) and (11.46), we have

$$-\omega^2 a_1 = -\frac{k+k'}{m}a_1 + \frac{k}{m}a_2, \tag{11.49}$$

$$-\omega^2 a_2 = \frac{k}{m}a_1 - \frac{k+k'}{m}a_2. \tag{11.50}$$

The condition is required on the matrix equation

$$\begin{pmatrix} \left(\omega^2 - \frac{k+k'}{m}\right) & \frac{k}{m} \\ \frac{k}{m} & \left(\omega^2 - \frac{k+k'}{m}\right) \end{pmatrix} \begin{pmatrix} a_1 \\ a_2 \end{pmatrix} = \begin{pmatrix} 0 \\ 0 \end{pmatrix} \tag{11.51}$$

to have non-trivial solutions other than $(a_1, a_2) = (0, 0)$. This requirement gives

$$\left(\omega^2 - \frac{k+k'}{m}\right)^2 - \left(\frac{k}{m}\right)^2 = 0. \tag{11.52}$$

By solving this equation with respect to $\omega$, we have

$$\omega = \sqrt{\frac{k+k'}{m} \pm \frac{k}{m}}, \tag{11.53}$$

where we ignored negative values $\omega < 0$ because they do not give rise to any independent solution. The two solutions are classified in the following way:

- The solution $\omega_1 = \sqrt{\frac{k'}{m}}$ gives $a_1 = a_2$. Since the displacements of both particles are in the same direction with the same magnitude, the center of mass oscillates while the relative motion is fixed. This solution is called the **center of mass motion**.
- The solution $\omega_2 = \sqrt{\frac{2k+k'}{m}}$ gives $a_1 = -a_2$. The two displacements are opposite to each other. The distance between the two particles oscillates while the center of mass does not move. This solution is called the **breathing mode**.

As is seen in Example 11.4, the frequency of the "center of mass motion" is the smallest where all the ions move the same direction. The smallest frequency gives the smallest phonon energy. If the spring constant is common ($k = k'$), it follows that $\omega_2 = \sqrt{3}\omega_1$.

Let us consider the problem of oscillations in terms of quantum mechanics. We quantize the Hamiltonian

$$H = \frac{1}{2m}p_1^2 + \frac{1}{2m}p_2^2 + \frac{k'}{2}z_1^2 + \frac{k}{2}(z_1 - z_2)^2 + \frac{k'}{2}z_2^2 \tag{11.54}$$

by replacing the momentum operators $p_1$ and $p_2$ by

$$p_1 \to \frac{\hbar}{i}\frac{\partial}{\partial z_1}, \tag{11.55}$$

$$p_2 \to \frac{\hbar}{i}\frac{\partial}{\partial z_2}. \tag{11.56}$$

Then, we obtain the following Schrödinger equation with the Hamiltonian (11.54),

$$\left[-\frac{\hbar^2}{2m}\left(\frac{\partial^2}{\partial z_1^2} + \frac{\partial^2}{\partial z_1^2}\right) + \frac{k'}{2}z_1^2 + \frac{k}{2}(z_1 - z_2)^2 + \frac{k'}{2}z_2^2\right]\psi(z_1, z_2) = E\psi(z_1, z_2). \tag{11.57}$$

It may look difficult to solve Eq. (11.57) because there is a cross term of $z_1$ and $z_2$, i.e., $z_1 z_2$. To avoid the cross term in Eq. (11.57), we transform the coordinates and introduce a new pair of coordinates

$$q_1 = \frac{1}{\sqrt{2}}(z_1 + z_2), \tag{11.58}$$

$$q_2 = \frac{1}{\sqrt{2}}(z_1 - z_2) \tag{11.59}$$

instead of $z_1$ and $z_2$. Note that the original coordinates are obtained from the new ones by

$$z_1 = \frac{1}{\sqrt{2}}(q_1 + q_2), \tag{11.60}$$

$$z_2 = \frac{1}{\sqrt{2}}(q_1 - q_2). \tag{11.61}$$

Since the differentiation operations are given by

$$\frac{\partial}{\partial z_1} = \frac{\partial q_1}{\partial z_1}\frac{\partial}{\partial q_1} + \frac{\partial q_2}{\partial z_1}\frac{\partial}{\partial q_2}$$
$$= \frac{1}{\sqrt{2}}\left(\frac{\partial}{\partial q_1} + \frac{\partial}{\partial q_2}\right), \tag{11.62}$$

$$\frac{\partial}{\partial z_2} = \frac{1}{\sqrt{2}}\left(\frac{\partial}{\partial q_1} - \frac{\partial}{\partial q_2}\right), \tag{11.63}$$

we have the following form of Schrödinger equation with respect to $q_1$ and $q_2$:

$$\left[-\frac{\hbar^2}{2m}\left(\frac{\partial^2}{\partial q_1^2}+\frac{\partial^2}{\partial q_2^2}\right)+\frac{k'}{2}q_1^2+\frac{1}{2}(2k+k')q_2^2\right]\psi(q_1,q_2)=E\psi(q_1,q_2). \tag{11.64}$$

Now no cross term of $q_1$ and $q_2$ is present, and we can separate the wave function and the energy into the following form,

$$\psi(q_1,q_2)=\psi_1(q_1)+\psi_2(q_2), \tag{11.65}$$

$$E=E_1+E_2 \tag{11.66}$$

and obtain two independent equations with respect to $q_1$ and $q_2$:

$$\left[-\frac{\hbar^2}{2m}\frac{\partial^2}{\partial q_1^2}+\frac{k'}{2}q_1^2\right]\psi_1(q_1)=E_1\psi_1(q_1), \tag{11.67}$$

$$\left[-\frac{\hbar^2}{2m}\frac{\partial^2}{\partial q_2^2}+\frac{1}{2}(2k+k')q_2^2\right]\psi_2(q_2)=E_2\psi_2(q_2). \tag{11.68}$$

Note that Eqs. (11.67) and (11.68) correspond to the two oscillations $\omega_1=\sqrt{k'/m}$ and $\omega_2=\sqrt{(2k+k')/m}$, respectively, which are exactly the same as those obtained in the classical mechanics problem. In fact, in general the normal coordinates $q_1$, $q_2$, ... are determined corresponding to the normal modes obtained by the diagonalization of the matrix in the classical mechanics problems, though we have treated the classical and the quantum mechanical problems as separate ones in the case illustrated above.

The two equations (11.67) and (11.68) are called the **harmonic oscillator problem**, often seen in the problems of quantum mechanics and their energies are known to be

$$E_1=\frac{1}{2}\hbar\omega_1,\ \frac{3}{2}\hbar\omega_1,\ \frac{5}{2}\hbar\omega_1,\ \ldots, \tag{11.69}$$

$$E_2=\frac{1}{2}\hbar\omega_2,\ \frac{3}{2}\hbar\omega_2,\ \frac{5}{2}\hbar\omega_2,\ \ldots. \tag{11.70}$$

The coupled Hamiltonian of the intrinsic motion in ions and the oscillation of the center of mass can be written as

$$H=\sum_i H_{\text{in}}(i)+\hbar\omega_1\left(a^\dagger a+\frac{1}{2}\right), \tag{11.71}$$

where the operators

$$a^\dagger=\frac{1}{\sqrt{2}}\left(\nu q_1-\nu^{-1}\frac{\partial}{\partial q_1}\right), \tag{11.72}$$

$$a=\frac{1}{\sqrt{2}}\left(\nu q_1+\nu^{-1}\frac{\partial}{\partial q_1}\right) \tag{11.73}$$

with $\nu = \sqrt{m\omega_1/\hbar}$, are called the creation and the annihilation operators respectively. The operator

$$N = a^\dagger a \tag{11.74}$$

is called the number operator whose eigenvalue corresponds to the quantum number to specify the excited states (phonon states) and have the values $n = \langle N \rangle = 0, 1, 2, \ldots$.

In the following sections, let us assume two electronic levels in each ion and the center of mass motion as the oscillation. Then, we have the energy

$$E = \sum_{i=1}^{2} E_{\text{in}}(i) + \sum_{n=0}^{\infty} \hbar\omega_1 \left(n + \frac{1}{2}\right). \tag{11.75}$$

### 11.2.3 *Computation*

Since the energy difference between levels in ions is of the order of 1 eV, its radiation has the wavelength $\approx 1\ \mu$m close to visible light (400 ~ 800nm). On the other hand, the radius of laser beam can be reduced to the order of its wavelength so that it is possible to control each ion separately if the ions are allocated at an interval of order 10 $\mu$m. Adopting the ground and the first excited states of each ion as a qubit, one can realize $n$ quantum bit computer by $n$ ions. One can also make use of the oscillation of the ion system as qubits.

In order to control the system, the **Rabi flopping** is used, i.e., the pulsed laser beam, having the frequency corresponding to the energy difference of two levels, induce the transition between qubit levels as the result of certain time application of laser beam. To understand the phenomena, let us introduce the following simple model. Firstly, the ground state and an excited state ($E = \hbar\omega_0$) are given by

$$|0\rangle = \begin{pmatrix} 1 \\ 0 \end{pmatrix}, \qquad |1\rangle = \begin{pmatrix} 0 \\ 1 \end{pmatrix}. \tag{11.76}$$

In the semi-classical model, we treat the oscillation of electromagnetic wave as a non-diagonal element of a matrix, and write it as $Ve^{i\omega t}$. The Hamiltonian is represented by a $2 \times 2$ matrix,

$$H = \hbar\omega_0 \begin{pmatrix} 0 & 0 \\ 0 & 1 \end{pmatrix} + V \begin{pmatrix} 0 & e^{+i\omega t} \\ e^{-i\omega t} & 0 \end{pmatrix}. \tag{11.77}$$

This can be modified as

$$\begin{aligned}
H &= -\hbar\omega_0 \begin{pmatrix} 1/2 & 0 \\ 0 & -1/2 \end{pmatrix} \\
&\quad + 2V\cos\omega t \begin{pmatrix} 0 & 1/2 \\ 1/2 & 0 \end{pmatrix} + 2V\sin\omega t \begin{pmatrix} 0 & i/2 \\ -i/2 & 0 \end{pmatrix} \\
&\quad + \frac{\hbar\omega_0}{2}\mathbf{I} \\
&= -\hbar\omega_0\, s_z + 2V(\cos\omega t\, s_x - \sin\omega t\, s_y) + \frac{\hbar\omega_0}{2}\mathbf{I}. \qquad (11.78)
\end{aligned}$$

One can see that Eq. (11.78) has the same form as Eq. (11.5) for NMR by the replacement,

$$\gamma B_0 \to \hbar\omega_0, \qquad (11.79)$$

$$\gamma B_1 \to -2V \qquad (11.80)$$

apart from a constant term. Though the matrices in Eq. (11.78) do not refer to the spin, the similar result to Eq. (11.16) for NMR could be derived in the same way. The logical operations are specified by the duration of beam time. In particular, the pulse of duration to cause the transitions,

$$|0\rangle \to |1\rangle \qquad (11.81)$$

$$|1\rangle \to |0\rangle \qquad (11.82)$$

corresponds to the rotation by the angle 180° and is called $\pi$ **pulse**, while the pulse corresponding to the rotation by 90° is called $\pi/2$ **pulse**. The Hadamard transformation is attained by the $\pi/2$ pulse apart from some phase factor. The $2\pi$ **pulse** does not change the logical value of any state, but it changes the overall sign of the state. Thus, one can implement Not-gate, Hadamard transformation and phase gate for the ion trap computer by the above procedure.

### 11.2.4 *Preparation of the initial states*

Since some trapped ions are expected to be in the excited states, we need to reset the state of the system of ions before starting computation. As the initial state, it will be reasonable to set the system to be "0", i.e., the ground state. When the system is in the excited state, it is known that it will decay spontaneously to the ground state emitting light. But when it is in a meta-stable state, it will take quite a long time before it decays. To overcome these circumstances, a technique called the **laser cooling**

**method** is employed for bringing the system to the ground state effectively. In this method, one gives some energy to the system by laser to excite the system to an auxiliary state with high excitation energy so that the ion spontaneously decays to lower states emitting large amount of energies. By repeating this process, one leads the system of ions finally to the ground state. After the system is initialized to be the ground state $|0\rangle$, one can attain to set the logical value $|1\rangle$ either for the excited state or some superposed states by applying light with frequency corresponding to the energy difference between $|0\rangle$ and $|1\rangle$ for proper duration of time.

### 11.2.5 *Reading results*

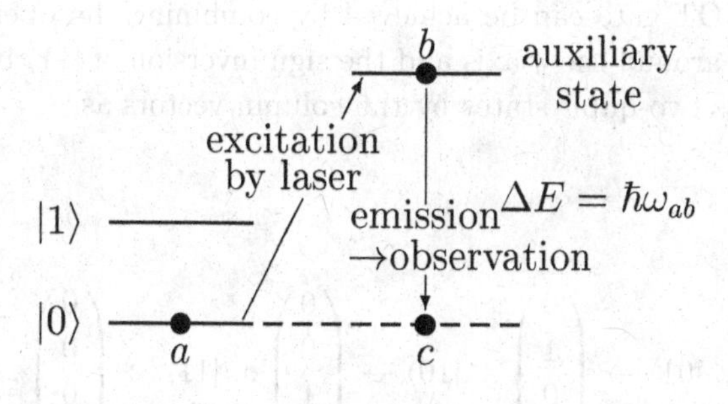


Fig. 11.15 Reading results from an ion trap.

When the computation is finished, we must read the result. If the system is in the excited state $|1\rangle$, we can recognize it in principle by the spontaneous decay to the ground state with emitted photons. The problem is that the decay time may be much long than that of the quantum computation. One can solve this problem by the following procedure. .

- First, one applies the laser beam of frequency corresponding to the energy difference between the ground state "$a$" and the auxiliary state "$b$", $\Delta E = \hbar\omega_{ab}$, as shown in Fig. 11.15.
- If the ion is in the ground state $|0\rangle$, the photon from laser will be absorbed, and the state of ion will be excited to the auxiliary state $b$. Then the ion will decay to the ground state emitting the photon with the energy $\Delta E = \hbar\omega_{ab}$ that will be observed by the detector.
- If the ion is in the excited state $|1\rangle$ orthogonal to $|0\rangle$, the absorption of laser will not occur, and the detector will observe nothing.
- Regarding the superposed states, one can observe the wave functions

by rotating the state before measurement.

The computation and reading results must be finished within the decoherence time limited by the spontaneous emission of light from excited states.

### 11.2.6 *Example of quantum gate*

#### 11.2.6.1 *Controlled-NOT*

The controlled-NOT gate can be achieved by combining the operation of the spin rotation around the $y$-axis and the sign inversion of $|11\rangle$ by the $2\pi$ pulse. We express two-qubit states by the column vectors as

$$|00\rangle \to \begin{pmatrix} 1 \\ 0 \\ 0 \\ 0 \end{pmatrix}, \quad |01\rangle \to \begin{pmatrix} 0 \\ 1 \\ 0 \\ 0 \end{pmatrix}, \quad |10\rangle \to \begin{pmatrix} 0 \\ 0 \\ 1 \\ 0 \end{pmatrix}, \quad |11\rangle \to \begin{pmatrix} 0 \\ 0 \\ 0 \\ 1 \end{pmatrix}. \tag{11.83}$$

The circuit of Fig. 11.16 acts as the controlled-NOT gate combining the

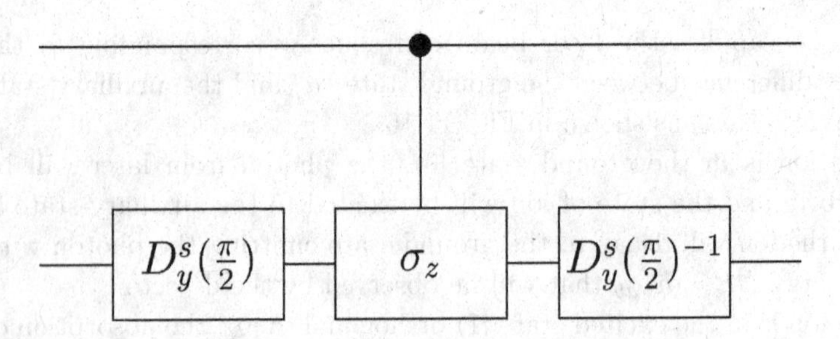


Fig. 11.16 Controlled-NOT implemented by elementary operations of ion trap. The sign inversion of $|1\rangle|1\rangle$ is expressed in the form of a controlled $\sigma_z$ gate.

spin rotation and the sign inversion:

$$D_y^s(\pi/2)\begin{pmatrix} 1 & 0, & 0, & 0 \\ 0 & 1 & 0 & 0 \\ 0 & 0 & 1 & 0 \\ 0 & 0 & 0 & -1 \end{pmatrix} D_y^s(-\pi/2)$$

$$= \begin{pmatrix} 1/\sqrt{2} & -1/\sqrt{2} & 0 & 0 \\ 1/\sqrt{2} & 1/\sqrt{2} & 0 & 0 \\ 0 & 0 & 1/\sqrt{2} & -1/\sqrt{2} \\ 0 & 0 & 1/\sqrt{2} & 1/\sqrt{2} \end{pmatrix} \begin{pmatrix} 1 & 0, & 0, & 0 \\ 0 & 1 & 0 & 0 \\ 0 & 0 & 1 & 0 \\ 0 & 0 & 0 & -1 \end{pmatrix}$$

$$\times \begin{pmatrix} 1/\sqrt{2} & 1/\sqrt{2} & 0 & 0 \\ -1/\sqrt{2} & 1/\sqrt{2} & 0 & 0 \\ 0 & 0 & 1/\sqrt{2} & 1/\sqrt{2} \\ 0 & 0 & -1/\sqrt{2} & 1/\sqrt{2} \end{pmatrix}$$

$$= \begin{pmatrix} 1 & 0, & 0, & 0 \\ 0 & 1 & 0 & 0 \\ 0 & 0 & 0 & 1 \\ 0 & 0 & 1 & 0 \end{pmatrix} = \text{Controlled-NOT}. \tag{11.84}$$

The operation of the $2\pi$ pulse being selectively to the state $|1\rangle|1\rangle$ can be implemented by the help of the auxiliary level which does not correspond to any qubit as shown in Fig. 11.17. As the $2\pi$ pulse, one applies the laser beam which resonates with the energy difference between the auxiliary level

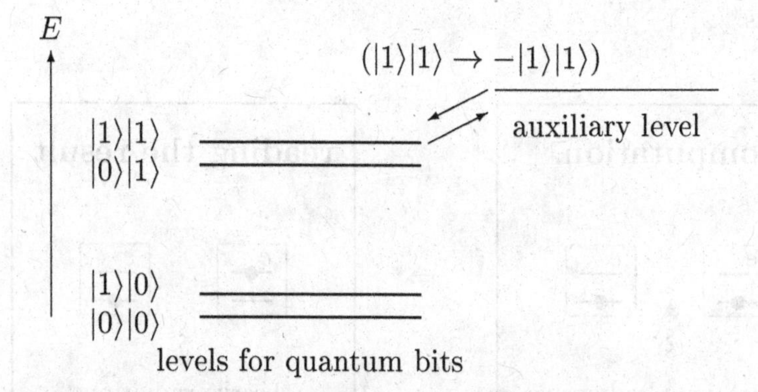


Fig. 11.17 Ion trap computer acts as the controlled-NOT, which has the level structure composed of the intrinsic excited state of an ion and the oscillation state. The auxiliary level is used to change the sign of the state $|1\rangle|1\rangle$. In the pair of kets, the left one corresponds to the oscillation state, while the right one is to the intrinsic electronic level.

and $|1\rangle|1\rangle$. Then, the state $|1\rangle|1\rangle$ is brought to the auxiliary level and decays to the original state with the inversed sign of the wave function. Since the resonance does not happen with other qubit states ($|0\rangle|0\rangle$, $|0\rangle|1\rangle$, $|1\rangle|0\rangle$), they are not changed by the laser beam. In this way, one can perform the selective control with the help of auxiliary levels.

## 11.3 Quantum dot computer

Quantum dot is also introduced to implement quantum computer. The idea is based on the usage of quantum dots as qubits.

### 11.3.1 *Principle*

It is possible to confine a single electron in a small space, whose size is comparable to the de Broglie wavelength (~1 nm), by fine processing of layers of semiconductors of different properties. The system obtained is called **quantum dot**. In the quantum dot computer shown in Fig. 11.18, the qubit is associated with the energy levels of the electron confined in such a space.

Let us assume that the electron confined in each box has two energy levels. If electrostatic field is applied, each level will have an electric dipole moment. If the direction of the electrostatic field is in the positive direction of the $x$-axis, as shown in Fig. 11.19, the wave functions will be shifted along the $x$-axis.

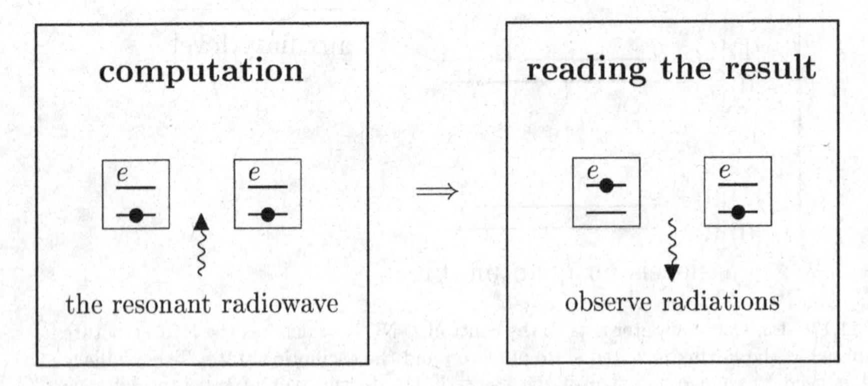


Fig. 11.18 Quantum dot computer. Each box represents a quantum dot with two level structure and its size is comparable to the de Broglie size.

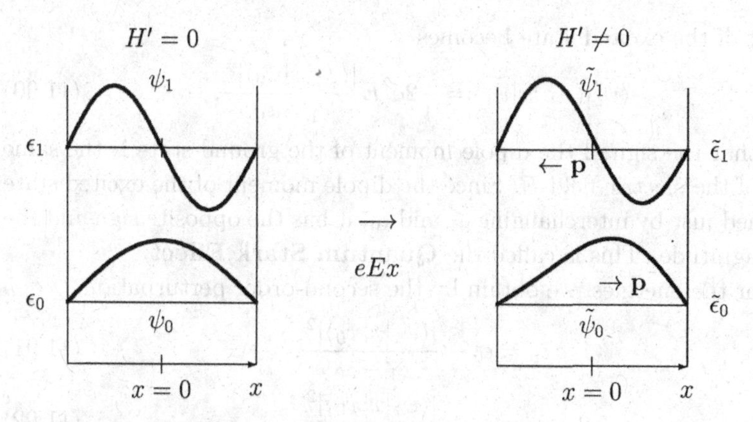


Fig. 11.19 Change of wave functions and energies of the ground state and the excited state induced by the electrostatic field $H' = -qEx = eEx$.

---

**Example 11.5** Consider a quantum dot in which an electron has two levels. Take $H_0$ as the Hamiltonian without external electrostatic field. A weak uniform electrostatic field $E$ is applied in the direction of the $x$-axis. Calculate the electric dipole moments of the ground state and the excited state induced by the field $E$. Show also that the ground state is polarized in the direction of the applied electrostatic field while the polarization of the excited state is in the opposite direction and has magnitude almost equal to that of the ground state.

**Solution** Let $|\psi_0\rangle$ and $|\psi_1\rangle$ be the ground state and the excited state of $H_0$, respectively:

$$H_0|\psi_0\rangle = \epsilon_0|\psi_0\rangle, \tag{11.85}$$

$$H_0|\psi_1\rangle = \epsilon_1|\psi_1\rangle. \tag{11.86}$$

The electrostatic field $E$ along the $x$-axis acting on the electron is represented by the potential $eEx$. When the perturbation $H' = -qEx = eEx$ is applied, the wave functions change into

$$|\tilde{\psi}_0\rangle = |\psi_0\rangle + \frac{\langle\psi_1|H'|\psi_0\rangle}{\epsilon_0 - \epsilon_1}|\psi_1\rangle, \tag{11.87}$$

$$|\tilde{\psi}_1\rangle = |\psi_1\rangle + \frac{\langle\psi_0|H'|\psi_1\rangle}{\epsilon_1 - \epsilon_0}|\psi_0\rangle \tag{11.88}$$

by the perturbation theory. As a result, the dipole moment of the ground state becomes

$$\langle\tilde{\psi}_0|(-e)x|\tilde{\psi}_0\rangle = 2e^2E\frac{|\langle\psi_1|x|\psi_0\rangle|^2}{\epsilon_1 - \epsilon_0}, \tag{11.89}$$

and that of the excited state becomes

$$\langle\tilde{\psi}_1|(-e)x|\tilde{\psi}_1\rangle = -2e^2 E \frac{|\langle\psi_1|x|\psi_0\rangle|^2}{\epsilon_1 - \epsilon_0}. \tag{11.90}$$

We see that the sign of the dipole moment of the ground state is the same as that of the electric field $E$. Since the dipole moment of the excited state is obtained just by interchanging $\epsilon_1$ and $\epsilon_0$, it has the opposite sign and the same magnitude. This is called the **Quantum Stark Effect**.

As for the energies, we obtain by the second-order perturbation,

$$\tilde{\epsilon}_0 = \epsilon_0 - \frac{|\langle\psi_1|x|\psi_0\rangle|^2}{\epsilon_1 - \epsilon_0}, \tag{11.91}$$

$$\tilde{\epsilon}_1 = \epsilon_1 + \frac{|\langle\psi_1|x|\psi_0\rangle|^2}{\epsilon_1 - \epsilon_0}. \tag{11.92}$$

$$\tag{11.93}$$

---

**Example 11.6** Calculate the electrostatic (Coulomb) potential energy between two separated charge distributions. Derive the formula for the potential energy expanded up to the order of dipole moment.

**Solution** First, we calculate the Coulomb potential by the first charge distribution (the charge distribution 1). Let $\mathbf{R}$ be the radial vector from the center of the first charge distribution to an arbitrary field point far away from the charge distribution. Coulomb With the vector $\mathbf{r_1}$ for the charge distribution 1 in Fig. 11.20 (a), the potential at the point $\mathbf{R}$ is calculated as

$$\phi_1(\mathbf{R}) = \frac{1}{4\pi\epsilon_0}\int \frac{\rho_1(\mathbf{r_1})}{|\mathbf{R} - \mathbf{r_1}|}\, d\mathbf{r_1}. \tag{11.94}$$

Since the radial vector $\mathbf{R}$ is much larger than the size of the charge distribution specified by $r_1$, i.e., $r_1 \ll R$ we can make the approximation up to the first order of $r_1$ (see Problem of this chapter);

$$\frac{1}{|\mathbf{R} - \mathbf{r_1}|} \approx \frac{1}{|\mathbf{R}|} + \frac{(\mathbf{r_1}\cdot\mathbf{R})}{|\mathbf{R}|^3}, \tag{11.95}$$

in Eq. (11.94). Then we obtain

$$\begin{aligned}\phi_1(\mathbf{R}) &= \frac{1}{4\pi\epsilon_0}\frac{1}{|\mathbf{R}|}\int \rho_1(\mathbf{r_1})\, d\mathbf{r_1} + \frac{1}{4\pi\epsilon_0}\frac{\mathbf{R}}{|\mathbf{R}|^3}\cdot\int \mathbf{r_1}\rho_1(\mathbf{r_1})\, d\mathbf{r_1} \\ &= \frac{1}{4\pi\epsilon_0}\frac{Q_1}{|\mathbf{R}|} + \frac{1}{4\pi\epsilon_0}\frac{(\mathbf{p_1}\cdot\mathbf{R})}{|\mathbf{R}|^3},\end{aligned} \tag{11.96}$$

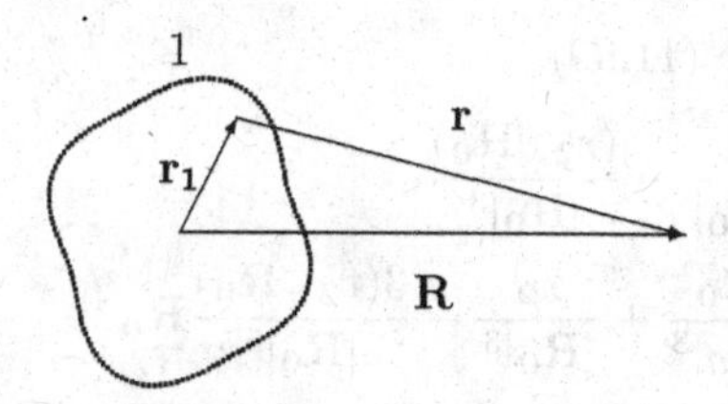


(a) Calculation of electrostatic potential by charge distribution 1

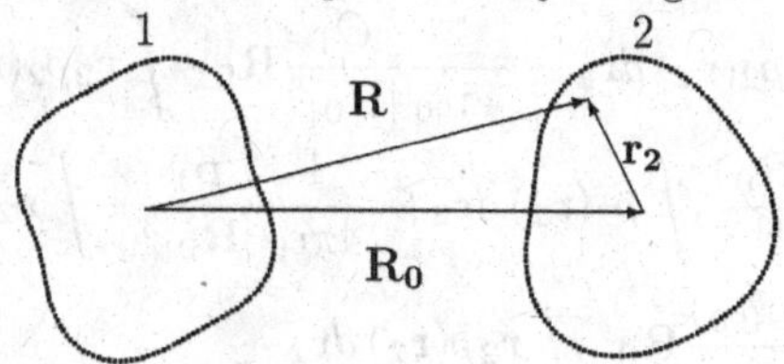


(b) Potential energy of charge distribution 2 in the potential by charge distribution 1

Fig. 11.20 Coulomb potential energy between two charge distributions.

where

$$Q_1 = \int \rho_1(\mathbf{r_1})\, d\mathbf{r_1}, \tag{11.97}$$

$$\mathbf{p_1} = \int \mathbf{r_1}\rho_1(\mathbf{r_1})\, d\mathbf{r_1} \tag{11.98}$$

are the total charge and the electric dipole moment of charge distribution 1, respectively.

Now, let us calculate the potential energy of the second charge distribution (the charge distribution 2) in the potential caused by the charge distribution 1. As shown in Fig. 11.20, let $\mathbf{R_0}$ be the vector from the center of charge distribution 1 to the center of charge distribution 2, and $\mathbf{r_2}$ be the vector from the center of charge distribution 2. We have

$$\begin{aligned} U &= \int \rho_2(\mathbf{r_2})\phi_1(\mathbf{R_0} + \mathbf{r_2})\, d\mathbf{r_2} \\ &= \frac{1}{4\pi\epsilon_0} Q_1 \int \frac{\rho_2(\mathbf{r_2})}{|\mathbf{R_0} + \mathbf{r_2}|}\, d\mathbf{r_2} + \frac{1}{4\pi\epsilon_0} \mathbf{p_1} \cdot \int \frac{\rho_2(\mathbf{r_2})(\mathbf{R_0} + \mathbf{r_2})}{|\mathbf{R_0} + \mathbf{r_2}|^3}\, d\mathbf{r_2}. \end{aligned} \tag{11.99}$$

Since $r_2 \ll R_0$, we can make the first order approximation with respect to

$r_2$ for the integrands in Eq. (11.99),

$$\frac{1}{|\mathbf{R_0}+\mathbf{r_2}|} \approx \frac{1}{|\mathbf{R_0}|} - \frac{(\mathbf{r_2}\cdot\mathbf{R_0})}{|\mathbf{R_0}|^3}, \tag{11.100}$$

$$\frac{\mathbf{R_0}+\mathbf{r_2}}{|\mathbf{R_0}+\mathbf{r_2}|^3} \approx \frac{\mathbf{R_0}}{|\mathbf{R_0}|^3} + \frac{\mathbf{r_2}}{|\mathbf{R_0}|^3} - \frac{3(\mathbf{r_2}\cdot\mathbf{R_0})}{|\mathbf{R_0}|^5}\mathbf{R_0}. \tag{11.101}$$

Then we have

$$\begin{aligned} U = & \frac{1}{4\pi\epsilon_0}\frac{Q_1}{|\mathbf{R_0}|}\int \rho_2(\mathbf{r_2})\,d\mathbf{r_2} - \frac{1}{4\pi\epsilon_0}\frac{Q_1}{|\mathbf{R_0}|^3}\mathbf{R_0}\cdot\int \mathbf{r_2}\rho_2(\mathbf{r_2})\,d\mathbf{r_2} \\ & + \frac{1}{4\pi\epsilon_0}\frac{(\mathbf{p_1}\cdot\mathbf{R_0})}{|\mathbf{R_0}|^3}\int \rho_2(\mathbf{r_2})\,d\mathbf{r_2} + \frac{1}{4\pi\epsilon_0}\frac{\mathbf{p_1}}{|\mathbf{R_0}|^3}\cdot\int \mathbf{r_2}\rho_2(\mathbf{r_2})\,d\mathbf{r_2} \\ & - \frac{1}{4\pi\epsilon_0}\frac{3(\mathbf{p_1}\cdot\mathbf{R_0})}{|\mathbf{R_0}|^5}\mathbf{R_0}\cdot\int \mathbf{r_2}\rho(\mathbf{r_2})\,d\mathbf{r_2} \\ = & \frac{1}{4\pi\epsilon_0}\frac{Q_1Q_2}{|\mathbf{R_0}|} + \frac{1}{4\pi\epsilon_0}\frac{\mathbf{p_1}\cdot\mathbf{p_2}}{|\mathbf{R_0}|^3} - \frac{1}{4\pi\epsilon_0}\frac{3(\mathbf{p_1}\cdot\mathbf{R_0})(\mathbf{p_2}\cdot\mathbf{R_0})}{|\mathbf{R_0}|^5} \\ & - \frac{1}{4\pi\epsilon_0}\frac{Q_1(\mathbf{p_2}\cdot\mathbf{R_0})}{|\mathbf{R_0}|^3} + \frac{1}{4\pi\epsilon_0}\frac{Q_2(\mathbf{p_1}\cdot\mathbf{R_0})}{|\mathbf{R_0}|^3}, \end{aligned} \tag{11.102}$$

where

$$Q_2 = \int \rho_2(\mathbf{r_2})\,d\mathbf{r_2}, \tag{11.103}$$

$$\mathbf{p_2} = \int \mathbf{r_2}\rho_2(\mathbf{r_2})d\mathbf{r_2} \tag{11.104}$$

are the total charge and the electric dipole moment of the charge distribution 2, respectively. The first term on the right-hand side of Eq. (11.102) comes from the interaction between the two charges, while the remaining terms involve the dipole moments. In particular, if the systems are neutral and the total charges vanish, $Q_1 = Q_2 = 0$, we have the dipole-dipole interaction

$$U = \frac{1}{4\pi\epsilon_0}\frac{\mathbf{p_1}\cdot\mathbf{p_2}}{(\mathbf{R_0})^3} - \frac{3}{4\pi\epsilon_0}\frac{(\mathbf{p_1}\cdot\mathbf{R_0})(\mathbf{p_2}\cdot\mathbf{R_0})}{(\mathbf{R_0})^5}. \tag{11.105}$$

If the both dipoles are perpendicular to the line connecting the two quantum dots, we have $\mathbf{p_1}\cdot\mathbf{R_0} = \mathbf{p_2}\cdot\mathbf{R_0} = 0$. In this case, only the first and the second terms in the r.h.s. of Eq. (11.102) remains finite.

---

We will show how to construct the controlled-NOT gate making use of the dipole moments of two quantum dots. In Fig. 11.21, we treat the first quantum dot as the control bit, while the second one as the target bit. We

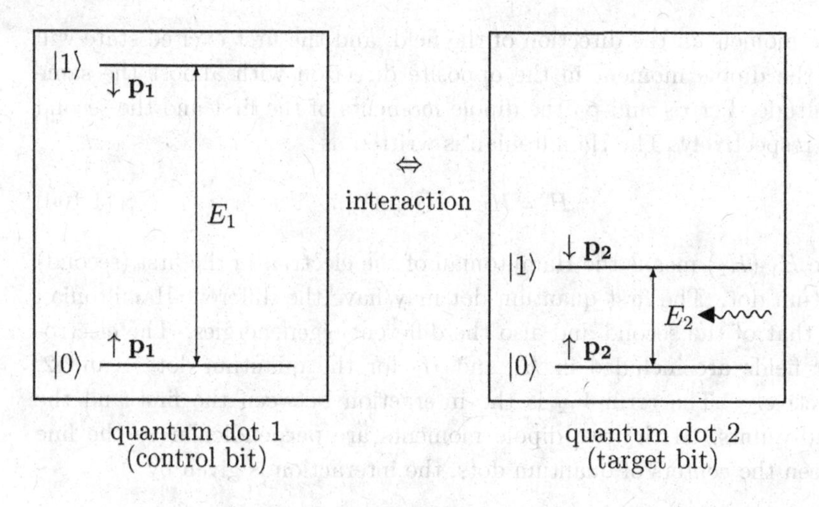


Fig. 11.21 Two quantum dots.

assign the logical value 0 to the ground state and the value 1 to the first excited state. If one applies the radiowave resonant to one of the excitation energy $E_1$ or $E_2$ of the qubit, the corresponding bit will get control by changing its logical value. However, as long as the two qubits are isolated from the other, they will not interact and have no influence on each other. If one applies the electrostatic field on the quantum dots along the $x$-axis, as shown in Example 11.6 and Fig. 11.21, the ground state will have the

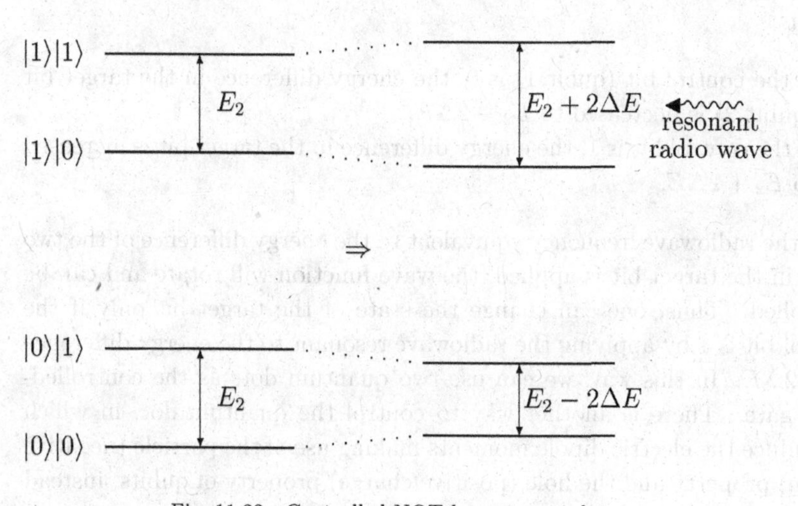


Fig. 11.22 Controlled-NOT by quantum dots.

dipole moment in the direction of the field, and the first excited state will have the dipole moment in the opposite direction with almost the same magnitude. Let $\mathbf{p_1}$ and $\mathbf{p_2}$ the dipole moments of the first and the second dots, respectively. The Hamiltonian is written as

$$H = H_1 + H_2 + V_{12}, \tag{11.106}$$

where $H_1$ ($H_2$) means the Hamiltonian of the electron in the first (second) quantum dot. The first quantum dot may have the different Hamiltonian from that of the second and also the different eigenenergies. The electrostatic fields are included in $H_1$ and $H_2$ for the quantum dots 1 and 2, respectively. The term $V_{12}$ is the interaction between the first and the second qubits. If the two dipole moments are perpendicular to the line between the centers of quantum dots, the interaction is given by

$$V_{12} = \frac{1}{4\pi\epsilon_0}\frac{\mathbf{p_1}\cdot\mathbf{p_2}}{R_0^3}. \tag{11.107}$$

from Eq. (11.105). If the two dipole moments are parallel, this interaction gives

$$\Delta E = \frac{1}{4\pi\epsilon_0}\frac{p_1 p_2}{R_0^3}, \tag{11.108}$$

while if they are anti-parallel, the contribution becomes $-\Delta E$. Therefore, the energy levels of the two quantum dots are shifted as shown in Fig. 11.22. That is,

- If the control bit (qubit 1) is 0, the energy difference in the target bit (qubit 2) is decreased to $E_2 - 2\Delta E$;
- If the control bit is 1, the energy difference in the target bit is increased to $E_2 + 2\Delta E$.

If the radiowave frequency equivalent to the energy difference of the two levels in the target bit is applied, the wave function will rotate and can be controlled. Thus, one can change the state of the target bit only if the control bit is 1 by applying the radiowave resonant to the energy difference $E_2 + 2\Delta E$. In this way, we can use two quantum dots as the controlled-NOT gate. There is another way to control the quantum dot, in which one induce the electric dipole moments making use of the particle (negative charge) property and the hole (positive charge) property of qubits, instead of the two particle nature quantum dots in Fig. 11.21.

**Summary of Chapter 11**

(1) Various physical systems are proposed and tested to implement quantum computers.
(2) The resonance between microwave and the nuclear spins in a static magnetic field is used to control qubits in the nuclear magnetic resonance (NMR) computer. The experiment of the factorization of 15 by NMR computer was conducted and its success was reported in 2001.
(3) In the ion trap computer, ions are trapped in space by oscillating electromagnetic field. Laser controls these ions.
(4) In the quantum dot computer, each electron is confined in a quantum dot, i.e., a very small space where the number of electronic energy levels is limited. There levels are adopted as qubits. The electrostatic potential and the radiowave control these qubits.

## 11.4 Problem

**[1]** There are two vectors $\mathbf{R} = (X, Y, Z)$ and $\mathbf{r} = (x, y, z)$. Assuming $R \gg r$, derive the following formula as the first-order expansion in $\mathbf{r}$,

$$\frac{1}{|\mathbf{R} - \mathbf{r}|} \approx \frac{1}{R} + \frac{(\mathbf{R} \cdot \mathbf{r})}{R^3}. \tag{11.109}$$

# Appendix A

# Basic Theory of Numbers

The numbers such as 1, 2, 3, ... are called natural numbers, and the set consisting of all natural numbers is written as **N**;

$$\mathbf{N} = \{1, 2, 3, \ldots\}. \tag{A.1}$$

Integers include natural numbers, 0, and negative numbers $-1, -2, -3, \ldots$. The set consisting of all integers is written as **Z**;

$$\mathbf{Z} = \{\ldots, -3, -2, -1, 0, 1, 2, 3, \ldots\}. \tag{A.2}$$

The set of all rational numbers is written as **Q**, the set of all real numbers is written as **R**, and the set of all complex numbers is written as **C**. The inclusion relation among them is

$$\mathbf{N} \subset \mathbf{Z} \subset \mathbf{Q} \subset \mathbf{R} \subset \mathbf{C}. \tag{A.3}$$

## A.1 Congruence

If the difference between two integers $a$ and $b$ is a multiple of $n$, $a$ and $b$ are called **congruent** with respect to **modulus** $n$. This relation is expressed as

$$a \equiv b \bmod n. \tag{A.4}$$

Equation (A.4) is called **congruence** (equation). This notation was proposed by a German mathematician Gauss. The congruence has the properties:

$$\text{if } a \equiv b \bmod n, \text{ then } b \equiv a \bmod n; \tag{A.5}$$
$$\text{if } a \equiv b \bmod n \text{ and } b \equiv c \bmod n, \text{ then } a \equiv c \bmod n. \tag{A.6}$$

The numbers which are congruent to a given number are identified to be in the same class as the given number. The numbers which are not congruent to each other, on the other hand, belong to different classes. Thus all integers can be classified by the congruence. Taking 2 as the modulus, for example, all integers are classified into two classes; one with even numbers and another with odd numbers.

If one performs divisions of all integers one by one by the integer $n$, namely modulo $n$, the residues, 0, 1, 2, ..., $n-1$ appear periodically. In this way, the set of all integers is separated into $n$ "**classes**" by the modulus $n$. If one knows an element $a$ of a class, all the elements that belong to this class can be expressed as

$$a + nt \qquad (t = 0, \pm 1, \pm 2, \pm 3, \ldots). \tag{A.7}$$

When one chooses one arbitrary integer from each class among $n$ different classes, the set of those $n$ integers is called the **residue system**. The class in which all elements are prime to $n$ is called the **residue class**. The number of the reduced classes modulo $n$ is called **Euler's (totient) function** $\varphi(n)$. The set of $\varphi(n)$-numbers chosen from all the reduced classes is called **reduced residue system**.

---

**Example** Classify integers modulo $n = 8$.

Table A.1 Classes of integers modulo $n = 8$.

| $c_0$ | $c_1$ | $c_2$ | $c_3$ | $c_4$ | $c_5$ | $c_6$ | $c_7$ |
|---|---|---|---|---|---|---|---|
| ⋮ | ⋮ | ⋮ | ⋮ | ⋮ | ⋮ | ⋮ | ⋮ |
| −16 | −15 | −14 | −13 | −12 | −11 | −10 | −9 |
| −8 | −7 | −6 | −5 | −4 | −3 | −2 | −1 |
| 0 | 1 | 2 | 3 | 4 | 5 | 6 | 7 |
| 8 | 9 | 10 | 11 | 12 | 13 | 14 | 15 |
| 16 | 17 | 18 | 19 | 20 | 21 | 22 | 23 |
| ⋮ | ⋮ | ⋮ | ⋮ | ⋮ | ⋮ | ⋮ | ⋮ |

**Solution** Dividing each integer by the modulus $n = 8$ successively, one classifies integers as shown in Table A.1. Denoting the residue class as $c_i$, one finds eight classes: $c_i = \{c_0, c_1, \ldots, c_7\}$. By choosing one element from each column of Table A.1, one obtains, for example, the following residue systems:

| −8 | 1 | 10 | −13 | −4 | 5 | 6 | 23 |
|---|---|---|---|---|---|---|---|
| 0 | 1 | 2 | 3 | 4 | 5 | 6 | 7 |

Taking $n$ as the modulus, the positive residues less than $n$ are called the least positive residues. In Table A.1, the least positive residue system is composed of

$$1, 2, 3, 4, 5, 6, 7. \tag{A.8}$$

There are four reduced classes with $n = 8$. Thus Euler's function is given as $\varphi(8) = 4$. For the reduced residue system, one can choose, for example,

| −7 | −13 | 5 | 15 |
|---|---|---|---|
| 1 | 3 | 5 | 7 . |

---

**Example A.1** Assume that the relations $a \equiv a'$ mod $m$ and $b \equiv b'$ mod $m$ hold. Show that the following equations hold for addition, subtraction and multiplication:

$$a \pm b \equiv (a' \pm b') \bmod m \tag{A.9}$$

$$ab \equiv (a'b') \bmod m. \tag{A.10}$$

Generally, under the conditions, $a \equiv a'$ mod $m$, $b \equiv b'$ mod $m$, $c \equiv c'$ mod $m$, ..., show that the equation

$$f(a, b, c, \ldots) \equiv f(a', b', c', \ldots) \bmod m \tag{A.11}$$

holds for a function $f(x, y, z, \ldots)$ if it is a polynomial of $x, y, z, \ldots$ with integer coefficients.

**Solution** From $a \equiv a'$ mod $m$ and $b \equiv b'$ mod $m$, $a - a'$ and $b - b'$ are multiples of $m$:

$$a = a' + \alpha m, \qquad b = b' + \beta m \qquad (\alpha, \beta \text{ an integer}). \tag{A.12}$$

Thus we have

$$a + b = a' + b' + (\alpha + \beta)m \tag{A.13}$$

$$a - b = a' - b' + (\alpha - \beta)m, \tag{A.14}$$

namely

$$a \pm b \equiv a' \pm b' \bmod m \tag{A.15}$$

for the addition and the subtraction. As for the multiplication, from

$$ab - a'b' = (a - a')b + a'(b - b') \tag{A.16}$$

$$= (\alpha b + \beta a')m, \tag{A.17}$$

it follows that

$$ab \equiv a'b' \bmod m. \tag{A.18}$$

From $Na = Na' + N\alpha m$ with an integer $N$, we have

$$Na \equiv Na' \bmod m. \tag{A.19}$$

From Eqs. (A.18) and (A.19), the relation

$$Na^p b^q c^r \cdots \equiv Na'^p b'^q c'^r \cdots \bmod m \tag{A.20}$$

holds for any powers of integers $p, q, r$. Since Eq. (A.20) holds for any number of terms, the relation

$$\sum_i N_i a^{pi} b^{qi} c^{ri} \cdots \equiv \sum_i N_i a'^{pi} b'^{qi} c'^{ri} \cdots \bmod m \tag{A.21}$$

also holds. Therefore, for any polynomial function $f(x, y, z, \ldots)$ with integer coefficients,

$$f(a, b, c, \ldots) \equiv f(a', b', c', \ldots) \bmod m. \tag{A.22}$$

---

---

**Example A.2** Assume

$$ac \equiv bc \bmod m. \tag{A.23}$$

Show

$$a \equiv b \bmod m \tag{A.24}$$

if $\gcd(c, m) = 1$. Also show

$$a \equiv b \bmod m' \tag{A.25}$$

if $\gcd(c, m) = d$, where $m' = m/d$.

**Solution** From Eq. (A.23), we have

$$ac = bc + \alpha m \qquad (\alpha \text{ is an integer}) \tag{A.26}$$

$$(a - b)c = \alpha m. \tag{A.27}$$

If $\gcd(c, m) = 1$, $(a - b)$ is a multiple of $m$ and we have

$$a \equiv b \bmod m. \tag{A.28}$$

In the case with $\gcd(c, m) = d$, $m = dm'$ and $c = dc'$. Then, we have

$$(a - b)dc' = \alpha dm', \qquad (a - b)c' = \alpha m'. \tag{A.29}$$

Thus $a - b$ is a multiple of $m'$. Therefore we prove

$$a \equiv b \bmod m' \qquad (m' = m/d). \tag{A.30}$$

---

## A.2 Euler's theorem (Fermat's little theorem)

If $\gcd(a, n) = 1$ for a natural number $n$ and an integer $a$, one obtains

$$a^{\varphi(n)} \equiv 1 \bmod n, \tag{A.31}$$

where $\varphi(n)$ is Euler's function. The above equality is called Euler's theorem. In particular, if $n$ is a prime number, Euler's function reduces to $\varphi(n) = n-1$ because it is the number of natural numbers which are less than $n$ and prime to $n$ at the same time. In this case, Eq. (A.31) is called Fermat's little theorem. Since Euler's function is given by the number of elements relatively prime to $n$ among the natural numbers less than $n$, it is the same as the number of elements of the reduced residue system modulo $n$. If $n = 12$, for example, the integers relatively prime to $n$ are

$$\{1, 5, 7, 11\}. \tag{A.32}$$

For any integer $k$, the integers given by

$$c_1 = 12k + 1 \tag{A.33}$$

$$c_5 = 12k + 5 \tag{A.34}$$

$$c_7 = 12k + 7 \tag{A.35}$$

$$c_{11} = 12k + 11 \tag{A.36}$$

are relatively prime to $n$ and compose the reduced residue class. The number of elements gives $\varphi(n = 12) = 4$. If we choose an element from each class, for example,

$$\{1, 5, 7, 11\},\ \{13, -7, 19, -1\},\ \{\cdots\}, \tag{A.37}$$

each set constitutes the reduced residue system. Let us pick up one of them $\{y_1, y_2, y_3, \ldots, y_l\}$ $(l = \varphi(n))$, multiply each by $a$ for which $\gcd(a, n) = 1$, and generate a set with $z_i$ where $ay_i \equiv z_i \bmod n$. It is shown that the resulted set: $\{z_1, z_2, z_3, \ldots, z_l\}$ $(l = \varphi(n))$, also constitutes the reduced residue system modulo $n$.

---

**Example A.3** Let

$$\{y_1, y_2, y_3, \ldots, y_l\} \quad (l = \varphi(n)) \tag{A.38}$$

be the reduced residue system modulo $n$. Show that the numbers obtained by multiplying the elements of this system by an integer $a$ relatively prime to $n$ also constitute the reduced residue system modulo $n$.

**Solution** Let us examine that the set $\{z_1, \ldots, z_l\}$ obtained as a result of congruence

$$ay_i \equiv z_i \bmod n \tag{A.39}$$

constitutes the reduced residue system, with the help of the relation $\gcd(a, n) = \gcd(y_i, n) = 1$. Suppose for two different elements $y_i \neq y_j$, that the relation,

$$ay_i \equiv ay_j \bmod n, \tag{A.40}$$

holds. Then, because $\gcd(a, n) = 1$, Eqs. (A.23) and (A.24) of Example A.2 gives

$$y_i \equiv y_j \bmod n, \tag{A.41}$$

which is in contradiction to the hypothesis that $y_i$ and $y_j$ are two different elements of the reduced residue system. Therefore $z_i$ and $z_j$ should be different elements of the residue system. At the same time, from $\gcd(a, n) = \gcd(y_i, n) = 1$, it follows that $\gcd(ay_i, n) = 1$. In this way, it has been shown that $\{z_1, \ldots, z_l\}$ are elements of the reduced residue system.

---

For example, by multiplying the reduced residue system $\{1, 5, 7, 11\}$ module 12 by $a = 5$, we have

$$\{a = 5, 5a = 25, 7a = 35, 11a = 55\} \overset{\text{mod }(12)}{\longrightarrow} \{5, 1, 11, 7\}, \tag{A.42}$$

namely, the resulted set is also the reduced residue system.

**Proof of Euler's theorem**

Let us consider two products,

$$P = y_1 y_2 \cdots y_l, \tag{A.43}$$

$$Q = ay_1 ay_2 \cdots ay_l = a^l P, \tag{A.44}$$

derived from the reduced residue system $\{y_1, y_2, \ldots, y_l\}$ of congruences modulo a natural number $n$. The integer $a$ satisfies $\gcd(n, a) = 1$. Because both $\{y_1, \ldots, y_l\}$ and $\{ay_1, \ldots, ay_l\}$ are reduced residue systems, it follows, from Example A.3 that for each $ay_i$ there exists unique solution $y_j$ that satisfies

$$ay_i \equiv y_j \bmod n. \tag{A.45}$$

Then, the set of $y_j$ constitute also the reduced residue system. The product of all elements is $P$. Therefore, by the theorem (A.10) for the products in congruence (see Example A.1), it follows that

$$Q \equiv a^l P \equiv P \bmod n. \tag{A.46}$$

Now, because $\gcd(P, n) = 1$, Eq. (A.46) divided by $P$ also holds, as proved in Example A.2. Thus we have

$$a^l \equiv a^{\varphi(n)} \equiv 1 \bmod n, \tag{A.47}$$

i.e., Euler's theorem has been proved. If $n$ is a prime number, in particular, (A.47) is called Fermat's little theorem.

---

**Example** Examples of Euler's theorem

i) Taking $n = 7, a = 2$, since $\varphi(7) = 6$, we have $a^{\varphi(n)} = 2^6 = 64$ and
$$2^6 \equiv 1 \bmod 7, \quad 2^6 - 1 = 63 = 7 \times 9.$$
ii) Taking $n = 8, a = 9$, since $\varphi(8) = 4$, we have $a^{\varphi(n)} = 9^4 = 6561$ and
$$9^4 \equiv 1 \bmod 8, \quad 9^4 - 1 = 6560 = 8 \times 820.$$
iii) Taking $n = 19, a = 2$, since $\varphi(19) = 18$, we have $a^{\varphi(n)} = 2^{18} = 262144$ and
$$2^{18} \equiv 1 \bmod 19, \quad 2^{18} - 1 = 262143 = 19 \times 13797.$$
iv) Taking $n = 11, a = 7$, since $\varphi(11) = 10$, we have $a^{\varphi(n)} = 7^{10} = 282475249$ and
$$7^{10} \equiv 1 \bmod 11, \quad 7^{10} - 1 = 282475248 = 11 \times 25679568.$$

---

## A.3 Euclid's algorithm

Euclid's algorithm is an algorithm to obtain the greatest common divisor (gcd) of two integers $a$ and $b$.

**Proposition** : Euclid's algorithm

**Step 1** : Let us assume $a > b$.
Divide $a$ by $b$. Let $r_1$ be the residue.

**Step 2** :
Divide $b$ by $r_1$. Let $r_2$ be the residue.

**Step 3** :
Divide $r_1$ by $r_2$. Let $r_3$ be the residue.

$\vdots$

**Step** $n+1$ :
Divide $r_{n-1}$ by $r_n$. Finish the computation when $r_{n+1} = 0$. The gcd is obtained as

$$\gcd(a, b) = r_n. \tag{A.48}$$

**Proof**

Taking $a$ and $b$ with the relation

$$a = qb + r \qquad (0 < r < b). \tag{A.49}$$

Assuming $d = \gcd(a, b)$, the integers $a$ and $b$ can be written as

$$a = da', \qquad b = db', \qquad a', b' \in \mathbf{Z}. \tag{A.50}$$

With $\gcd(b, r) = f$, $b$ and $r$ can be written as

$$b = fb'', \qquad r = fr'', \qquad b'', r'' \in \mathbf{Z}. \tag{A.51}$$

From Eqs. (A.49) and (A.51), we have

$$a = qfb'' + fr'' = f(qb'' + r''). \tag{A.52}$$

Therefore $f$ is a common divisor of $a$ and $b$. Because $d$ is the gcd of $a$ and $b$, $f \leq d$. On the other hand, from (A.49), we have

$$r = a - qb = da' - qdb' = d(a' - qb'), \tag{A.53}$$

namely, $d$ is a common divisor of $b$ and $r$. Because $f$ is the gcd of $b$ and $r$, $d \leq f$. Hence $d = f$ has been proved, i.e.,

$$\gcd(a, b) = \gcd(b, r) \tag{A.54}$$

Euclid's algorithm follows the steps described as

$$a = q_1 b + r_1 \qquad (b > r_1) \tag{A.55}$$
$$b = q_2 r_1 + r_2 \qquad (r_1 > r_2) \tag{A.56}$$
$$r_1 = q_3 r_2 + r_3 \qquad (r_2 > r_3) \tag{A.57}$$
$$\vdots$$
$$r_{n-1} = q_{n+1} r_n + r_{n+1} \qquad (r_n > r_{n+1} = 0). \tag{A.58}$$

Note that the $(n+1)$th step gives $\gcd(r_{n-1}, r_n) = r_n$. Since it follows from Eq. (A.54) that

$$\gcd(a,b) = \gcd(b, r_1) = \gcd(r_1, r_2) = \cdots = \gcd(r_{n-1}, r_n) = r_n, \tag{A.59}$$

the gcd of $a$ and $b$ is obtained as

$$\gcd(a,b) = r_n \tag{A.60}$$

**QED**

---

**Example A.4** Obtain the gcd of 195 and 143.

**Solution** Euclid's algorithm gives the steps,

$$195 = 1 \times 143 + 52,$$
$$143 = 2 \times 52 + 39,$$
$$52 = 1 \times 39 + 13,$$
$$39 = 3 \times 13 + 0.$$

Thus we end up with $\gcd(195, 143) = 13$ . The least common multiplier can be also obtained as

$$195 \times 143 / \gcd(195, 143) = 2145.$$

---

## A.4 Diophantus equation (Indeterminate equation)

Problems of obtaining integer solutions of the equations with integer coefficients for two or more unknowns are called indeterminate equations because they have infinite numbers of integer solutions if any solutions exist. The solution of linear indeterminate equations was given in an old book "Arithmetica" by Diophantus who flourished in Alexandria around A. D. 300. After him, the indeterminate equations with integer coefficients are called Diophantus equations.

Let us consider a problem of obtaining integer solutions of the linear indeterminate equation for $x$ and $y$,

$$ax + by = d, \qquad ab \neq 0, \tag{A.61}$$

with integer coefficients $a, b$ and $d$. In particular, if $d$ is $\gcd(a, b)$ or a multiple of it, the solutions can be obtained by using Euclid's algorithm.

---

**Example A.5** Obtain the indeterminate solutions of

$$195x + 143y = 13. \tag{A.62}$$

**Solution** Let us follow the steps of Euclid's algorithm in Example A.4 upside down:

$$13 = 52 - 1 \times 39. \tag{A.63}$$

By substituting the previous step $39 = 143 - 2 \times 52$, we have

$$13 = 52 - 1 \times (143 - 2 \times 52) = 3 \times 52 - 1 \times 143. \tag{A.64}$$

Next, we substitute $52 = 195 - 143$ to have

$$\begin{aligned} 13 = & \quad 3 \times (195 - 143) - 143 \\ = & \quad 3 \times 195 - 4 \times 143. \end{aligned} \tag{A.65}$$

In this way, we can identify that the equation $195x + 143y = 13$ has a solution $x = 3$ and $y = -4$. The general solutions $x$ and $y$ is expressed as

$$x = 3 + t_1 k, \qquad y = -4 + t_2 k, \qquad k \in \mathbf{Z}. \tag{A.66}$$

Substitute Eq. (A.66) in Eq. (A.62), we have

$$195 \times (3 + t_1 k) + 143(-4 + t_2 k) = 13, \tag{A.67}$$

and

$$(195t_1 + 143t_2) = 0. \tag{A.68}$$

Dividing Eq. (A.68) by $\gcd(195, 143) = 13$, we finally obtain

$$15t_1 + 11t_2 = 0. \tag{A.69}$$

If we choose $t_1 = 11$ and $t_2 = -15$, the general solution of Eq. (A.62) is given by

$$x = 3 + 11k, \qquad y = -4 - 15k \qquad k \in \mathbf{Z}. \tag{A.70}$$

---

If $d = \gcd(a, b)$, the general solution of the Diophantus equation (A.61) is given as

$$x = x_1 + b'k, \qquad y = y_1 - a'k \tag{A.71}$$

where $a' = \frac{a}{d}$, $b' = \frac{b}{d}$, and the pair $x_1$ and $y_1$ gives one of the solutions. In general, the necessary and sufficient condition for solving the linear indeterminate equation

$$ax + by = d \tag{A.72}$$

is that $d$ is a multiple of $\gcd(a, b)$.

## A.5 Chinese remainder theorem

The Chinese remainder theorem can be considered as a kind of Euclid's algorithm. Since the solution was found in an old Chinese book "Sun Zi suanjing" written around A. D. 300, the theorem is called the Chinese remainder.

**Proposition** : Suppose the system of equations

$$\begin{aligned} x &\equiv a_1 \bmod m_1 \\ x &\equiv a_2 \bmod m_2 \\ &\vdots \\ &\vdots \\ x &\equiv a_n \bmod m_n \end{aligned} \tag{A.73}$$

with $n$ natural numbers $m_i$ $(i = 1, \cdots, n)$ relatively prime to each other $(\gcd(m_i, m_j) = 1$ if $i \neq j)$, it has always solutions $x$ and all of them are congruent modulo $M = m_1 m_2 \cdots m_n$.

**Proof** : First, define $M_i = M/m_i$. Noting $\gcd(M_i, m_i) = 1$, the existence of $N_i$ satisfying

$$M_i N_i \equiv 1 \bmod m_i \tag{A.74}$$

can be shown in the following way with the help of the Diophantus equations. Introducing an integer $b$, let us express (A.74) as

$$M_i N_i - m_i b = 1. \tag{A.75}$$

This can be regarded as a linear indeterminate equation with respect to unknowns $N_i$ and $b$. Since the right-hand side is divisible by $\gcd(M_i, m_i)$, Eq. (A.75) has a solution, as was shown in Section A.1.4. Let us define

$$x = \sum_{i=1}^{n} a_i M_i N_i. \tag{A.76}$$

Because $M_i N_i \equiv 1 \bmod m_i$, and $M_j N_j = 0 \bmod m_i$ for $i \neq j$, it follows from the theorem regarding the addition of congruence equations that

$$x \equiv \sum_{i=1}^{n} a_i M_i N_i \equiv a_i \bmod m_i. \tag{A.77}$$

Thus $x$ defined as (A.76) is shown to be a solution of the system of equations (A.73).

Next, we prove the uniqueness of the solution modulo $M$. Let us assume that $x$ and $x'$ are solutions of the system of equations (A.73). Then, the difference satisfies

$$x - x' \equiv 0 \bmod m_i \qquad (i = 1, 2, \cdots, n) \tag{A.78}$$

for all $m_i$. Since $\gcd(m_i, m_j) = 1 \ (i \neq j)$, the equation

$$x - x' \equiv 0 \bmod M \tag{A.79}$$

holds also for $M = m_1 m_2 \cdots m_n$. Therefore it follows that

$$x \equiv x' \bmod M, \tag{A.80}$$

showing that the two solutions $x$ and $x'$ are congruent modulo $M$. **QED**

---

**Example A.6** (Multiplicativity of Euler's functions)

Suppose that two natural numbers $m$ and $n$ are relatively prime. Show that Euler's function of the product $\varphi(mn)$ can be calculated as the product $\varphi(mn) = \varphi(m)\varphi(n)$ of the two functions $\varphi(m)$ and $\varphi(n)$.

**Solution** Note that the value of $\varphi(mn)$ is the number of natural numbers which are less than $M = mn$ and relatively prime to $M$. Now, let $x = \{x_1, x_2, \cdots, x_{\varphi(m)}\}$ and $y = \{y_1, y_2, \cdots, y_{\varphi(n)}\}$ be the reduced residue system modulo $m$ and $n$, respectively. Out of $\varphi(m)\varphi(n)$–th kind of different combinations of $x, y$, corresponding to all pairs of elements taken from $x$ and $y$, there is an integer $z$ that satisfies

$$z \equiv x_i \bmod m$$
$$z \equiv y_j \bmod n.$$

Then the integer $z$ is congruent modulo $M$ (Chinese remainder theorem). Here naturally $z$ is relatively prime to $M$, i.e., $\gcd(z, M) = 1$. Conversely, if $\gcd(z, M) = 1$, then $\gcd(z, m) = 1$ and $\gcd(z, n) = 1$, and therefore the pair $(x_i, y_j)$ of integers that satisfy

$$z = x_i \bmod m$$
$$z = y_j \bmod n$$

are uniquely determined. In other words, one-to-one correspondence holds between a number $z$ of the reduced residue system of $M$ and a pair $(x_i, y_i)$. In this way, the equality $\varphi(M) = \varphi(mn) = \varphi(m)\varphi(n)$ has been proved.

---

**Example** Let us study the multiplicativity of Euler's function with $m = 3$ and $n = 7$. This case has $M = mn = 21$. From $\varphi(m) = 2$ and $\varphi(n) = 6$, let us write their respective reduced residue systems as

$$x = \{1, 2\}, \quad y = \{1, 2, 3, 4, 5, 6\}.$$

Then the set $z$ is generated by the combinations of the elements from $x$ and $y$ by Chinese remainder theorem as shown in Table A.2. Arranging the obtained elements in the increasing order, we have

$$z = \{1, 2, 4, 5, 8, 10, 11, 13, 16, 17, 19, 20\}.$$

We see that $\varphi(M) = 12 = 2 \cdot 6 = \varphi(m)\varphi(n)$ holds.

---

Table A.2 The reduced residue systems $x, y, z$.

| $x$ | 1 | 1 | 1 | 1 | 1 | 1 | 2 | 2 | 2 | 2 | 2 | 2 |
|---|---|---|---|---|---|---|---|---|---|---|---|---|
| $y$ | 1 | 2 | 3 | 4 | 5 | 6 | 1 | 2 | 3 | 4 | 5 | 6 |
| $z$ | 1 | 16 | 10 | 4 | 19 | 13 | 8 | 2 | 17 | 11 | 5 | 20 |

## A.6 Continued fractions

The continued fraction is defined as

$$a_0 + \cfrac{1}{a_1 + \cfrac{1}{a_2 + \cfrac{1}{a_3 + \cfrac{1}{a_4 + \cfrac{1}{\cdots + \cfrac{1}{a_N}}}}}} \tag{A.81}$$

where $a_0, \ldots, a_N$ are positive integers. Equation (A.81) is also written as

$$[a_0, \ldots, a_N]. \tag{A.82}$$

The number $[a_0, \ldots, a_n]$ with $n < N$ is called the $n-$th order convergent of $[a_0, \ldots, a_N]$. The $n-$th order convergent is written as a fraction $p_n/q_n$, where the integers $p_n$ and $q_n$ can be obtained in the following procedure: To start, noting the relation with the fractions $[a_0]$ and $[a_0, a_1]$

$$\begin{aligned} p_0 &= a_0, \qquad q_0 = 1 \\ \tfrac{p_1}{q_1} &= a_0 + \tfrac{1}{a_1} = \tfrac{a_0 a_1 + 1}{a_1}, \end{aligned} \tag{A.83}$$

we have

$$\begin{aligned} p_1 &= a_0 a_1 + 1 = a_1 p_0 + 1, \\ q_1 &= a_1 = a_1 q_0. \end{aligned} \tag{A.84}$$

Repeating the same procedure, we have a recursion formula for $p_n$ and $q_n$:

$$\begin{aligned} p_1 &= a_1 p_0 + 1 \cdots p_n = a_n p_{n-1} + p_{n-2} \\ q_1 &= a_1 q_0 \qquad \cdots q_n = a_n q_{n-1} + q_{n-2}. \end{aligned} \tag{A.85}$$

Taking the following procedure, any positive rational number $x$ is expressed in the form of continued fraction. The expression $\lfloor x \rfloor$ means the greatest integer that is less than or equal to $x$, while $\lceil x \rceil$ means the least integer that is greater or equal to $x$.

$$\begin{aligned} a_0 &= \lfloor x \rfloor \\ x &= a_0 + \xi_0 \qquad 0 \leq \xi_0 < 1 \end{aligned} \tag{A.86}$$

- If $\xi_0 \neq 0$, then take $a_1 = \lfloor \frac{1}{\xi_0} \rfloor$ and

$$\frac{1}{\xi_0} = a_1 + \xi_1 \qquad 0 \leq \xi_1 < 1$$

- If $\xi_1 \neq 0$, then take $a_2 = \lfloor \frac{1}{\xi_1} \rfloor$ and

$$\frac{1}{\xi_1} = a_1 + \xi_2 \qquad 0 \leq \xi_2 < 1$$

Repeat the similar processes until $\xi_n = 0$.

---

**Example A.7** Express $x = \frac{19}{11}$ as a continued fraction.

**Solution**

$$\begin{array}{ll} a_0 = 1 & \xi_0 = \frac{19}{11} - 1 = \frac{8}{11} \\ a_1 = \lfloor \frac{11}{8} \rfloor = 1 & \xi_1 = \frac{11}{8} - 1 = \frac{3}{8} \\ a_2 = \lfloor \frac{8}{3} \rfloor = 2 & \xi_2 = \frac{8}{3} - 2 = \frac{2}{3} \\ a_3 = \lfloor \frac{3}{2} \rfloor = 1 & \xi_3 = \frac{3}{2} - 1 = \frac{1}{2} \\ a_4 = 2 & \end{array}$$

$$\begin{aligned} x &= 1 + \cfrac{1}{1 + \cfrac{1}{2 + \cfrac{1}{1 + \cfrac{1}{2}}}} \\ &= 1 + \cfrac{1}{1 + \cfrac{1}{2 + \cfrac{2}{3}}} \\ &= 1 + \cfrac{1}{1 + \cfrac{3}{8}} \\ &= 1 + \frac{8}{11} \\ &= \frac{19}{11} \end{aligned}$$

---

---

**Example A.8** Derive the following equation regarding the $2 \times 2$ matrix

$$\begin{vmatrix} p_n & p_{n-1} \\ q_n & q_{n-1} \end{vmatrix} = (-1)^{n-1}$$

where $p_n/q_n$ and $p_{n-1}/q_{n-1}$ are the $n$−th and $(n-1)$−th convergents of a rational number, respectively, defined by (A.85).

**Solution** From the recursion formula for $p_n$, we have

$$\begin{vmatrix} p_n & p_{n-1} \\ q_n & q_{n-1} \end{vmatrix} = \begin{vmatrix} p_{n-1}a_{n-1} + p_{n-2} & p_{n-1} \\ q_{n-1}a_{n-1} + q_{n-2} & q_{n-1} \end{vmatrix}$$

$$= \begin{vmatrix} p_{n-2} & p_{n-1} \\ q_{n-2} & q_{n-1} \end{vmatrix} = - \begin{vmatrix} p_{n-1} & p_{n-2} \\ q_{n-1} & q_{n-2} \end{vmatrix}.$$

By repeating this procedure $n-1$ times, we have

$$\begin{vmatrix} p_n & p_{n-1} \\ q_n & q_{n-1} \end{vmatrix} = (-1)^{n-1} \begin{vmatrix} p_1 & p_0 \\ q_1 & q_0 \end{vmatrix} = (-1)^{n-1} \begin{vmatrix} a_1a_0 + 1 & a_0 \\ a_1 & 1 \end{vmatrix}$$

$$= (-1)^{n-1} \begin{vmatrix} 1 & a_0 \\ 0 & 1 \end{vmatrix} = (-1)^{n-1}.$$

---

**Proposition** If two positive integers $p$ and $q$ satisfy the inequality:

$$\left| \frac{p}{q} - x \right| \leq \frac{1}{2q^2} \tag{A.87}$$

with a rational number $x$, the value of $p/q$ agrees with a convergent of the continued fraction of $x$.

**Proof** Let $[a_0, \cdots, a_n]$ be the continued fraction of $p/q$. Let us write (A.87) as

$$x = \frac{p}{q} + \frac{\delta}{2q^2}$$

where $0 < |\delta| \leq 1$. By introducing $\lambda$ as

$$\lambda = 2\frac{q_n p_{n-1} - p_n q_{n-1}}{\delta} - \frac{q_{n-1}}{q_n},$$

where $p_n/q_n$ and $p_{n-1}/q_{n-1}$ are the convergents of $p/q$ defined by (A.85), we have

$$\frac{\lambda p_n + p_{n-1}}{\lambda q_n + q_{n-1}} = \frac{2p_n q_n(q_n p_{n-1} - p_n q_{n-1}) + \delta(q_n p_{n-1} - p_n q_{n-1})}{2q_n^2(q_n p_{n-1} - p_n q_{n-1})}$$

$$= \frac{p_n}{q_n} + \frac{\delta}{2q_n^2} = x.$$

It means that $x$ is expressed in the form of a continued fraction: $x = [a_0, \cdots, a_n, \lambda]$. Firstly we consider the case $\delta > 0$. If $n$ is even, we have $p_n q_{n-1} - q_n p_{n-1} = -1$ from Example A.8. Therefore,

$$\lambda = \frac{2}{\delta} - \frac{q_{n-1}}{q_n} > 2 - 1 = 1,$$

where we use the equation $q_{n-1}/q_n = 1/a_n < 1$. Since $\lambda$ turns out to be a rational number greater than unity, it can be expressed as $\lambda = [b_0, \cdots, b_m]$. Hence, we have $x = [a_0, \cdots, a_n, b_0, \cdots, b_m]$, namely, $p/q = [a_0, \cdots, a_n]$ agrees with the $n$th convergent of $x$. If $n$ is odd, we express the continued fraction of $p/q$ as $[a_0, \cdots, a_n - 1, 1]$. We define $\lambda$ with respect to the convergents derived from this form. Then the same argument applies because $n + 1$ is even. We thus have the continued-fraction of $x$ as $[a_0, \cdots, a_n - 1, 1, b_0, \cdots, b_m]$, while $p/q$ is its $(n+1)$th convergent: $[a_0, \cdots, a_n - 1, 1]$.

The case with $\delta < 0$ can be treated in a similar way with an odd $n$. For odd n, the argument is the same as even n with $\delta > 0$. For even n, the argument can be taken as the same with odd n with $\delta > 0$. **QED**

The case with an odd $n$ can be converted into an even-$n$ case by expressing the continued-fraction form of $p/q$ as $[a_0, \cdots, a_n - 1, 1]$. The case with $\delta < 0$ can be treated in a similar way.

# Appendix B

# Solutions of Problems

## [Chapter 2]

[1] a) yes. b) no. c) no. d) yes.
[2] Eigenstates are

$$e^{ax} \quad \text{for} \quad \frac{d}{dx},$$

$$e^{ax} \quad \text{and} \quad \sin x \quad \text{for} \quad \frac{d^2}{dx^2}.$$

[3] Since $a_1$ and $a_2$ are commutable with $H$ and do not depend on time,

$$\begin{aligned} i\hbar\frac{\partial}{\partial t}\Psi &= i\hbar\frac{\partial}{\partial t}(a_1\psi_1 + a_2\psi_2) = a_1 i\hbar\frac{\partial}{\partial t}\psi_1 + a_2 i\hbar\frac{\partial}{\partial t}\psi_2 \\ &= a_1 H\psi_1 + a_2 H\psi_2 = H(a_1\psi_1 + a_2\psi_2) = H\Psi \end{aligned}$$

[4] a) $< \Psi|p|\Psi >= \frac{1}{2}\left(< \psi_p|p|\psi_p > + < \psi_{-p}|p|\psi_{-p} >\right) = \frac{1}{2}(k-k) = 0$
b) $< \Psi|\frac{p^2}{2m}|\Psi >= \frac{1}{2}\left(\frac{k^2}{2m} + \frac{(-k)^2}{2m}\right) = \frac{k^2}{2m}$

## [Chapter 3]

[1] From Eqs. (3.49) and (3.50), we have

$$|\updownarrow\rangle = \frac{1}{\sqrt{2}}\{|R\rangle + |L\rangle\}, \tag{B.1}$$

$$|\leftrightarrow\rangle = -\frac{i}{\sqrt{2}}\{|R\rangle - |L\rangle\}. \tag{B.2}$$

Then we obtain

$$|\Psi\rangle_{12} = \frac{1}{\sqrt{2}}\{|R\rangle_1|R\rangle_2 + |L\rangle_1|L\rangle_2\}. \tag{B.3}$$

**[2]** Let $|\uparrow_y\rangle$ and $|\downarrow_y\rangle$ be the eigenstates of $s_y$ with eigenvalues $+\frac{1}{2}$ and $-\frac{1}{2}$, respectively. For the first eigenvector

$$|\uparrow_y\rangle = c|\uparrow_z\rangle + d|\downarrow_z\rangle = \begin{pmatrix} c \\ d \end{pmatrix}, \tag{B.4}$$

the normalization condition requires

$$|c|^2 + |d|^2 = 1. \tag{B.5}$$

From

$$\sigma_y|\uparrow_y\rangle = \begin{pmatrix} 0 & -i \\ i & 0 \end{pmatrix}\begin{pmatrix} c \\ d \end{pmatrix} = \begin{pmatrix} -id \\ ic \end{pmatrix} = \begin{pmatrix} c \\ d \end{pmatrix}, \tag{B.6}$$

we have

$$d = ic. \tag{B.7}$$

Choosing the overall phase so that $c$ becomes a positive number, we have

$$|\uparrow_y\rangle = \frac{1}{\sqrt{2}}\begin{pmatrix} 1 \\ i \end{pmatrix} = \frac{1}{\sqrt{2}}(|\uparrow_z\rangle + i|\downarrow_z\rangle). \tag{B.8}$$

As for $|\downarrow_y\rangle$, from

$$\sigma_y|\downarrow_y\rangle = \begin{pmatrix} 0 & -i \\ i & 0 \end{pmatrix}\begin{pmatrix} c' \\ d' \end{pmatrix} = \begin{pmatrix} -id' \\ ic' \end{pmatrix} = -\begin{pmatrix} c' \\ d' \end{pmatrix}, \tag{B.9}$$

we have $d' = -ic'$, therefore

$$|\downarrow_y\rangle = \frac{1}{\sqrt{2}}\begin{pmatrix} 1 \\ -i \end{pmatrix} = \frac{1}{\sqrt{2}}(|\uparrow_z\rangle - i|\downarrow_z\rangle). \tag{B.10}$$

**[3]** We have a commutation relation;

$$\begin{aligned}
[\mathbf{s}^2, s_z] &= [s_x^2, s_z] + [s_y^2, s_z] + [s_z^2, s_z] \\
&= s_x[s_x, s_z] + [s_x, s_z]s_x + s_y[s_y, s_z] + [s_y, s_z]s_y \\
&= s_x\left(-\frac{i}{2}s_y\right) + \left(-\frac{i}{2}s_y\right)s_x + s_y\frac{i}{2}s_x + \frac{i}{2}s_x s_y = 0.
\end{aligned} \tag{B.11}$$

Thus the commutability of $\mathbf{s}^2$ and $s_z$ is shown. The equality $[\mathbf{s}^2, s_x] = [\mathbf{s}^2, s_y] = 0$ is shown in the same way. In deriving Eq. (B.11), the relation:

$$\begin{aligned}
[s_i^2, s_j] &= s_i^2 s_j - s_j s_i^2 \\
&= s_i(s_i s_j - s_j s_i) + (s_i s_j - s_j s_i)s_i \\
&= s_i[s_i, s_j] + [s_i, s_j]s_i.
\end{aligned} \tag{B.12}$$

has been used.

[4] Equality (3.19) is shown as

$$\begin{aligned}
[\sigma_x, \sigma_y] &= [(|\uparrow\rangle\langle\downarrow| + |\downarrow\rangle\langle\uparrow|), -i(|\uparrow\rangle\langle\downarrow| - |\downarrow\rangle\langle\uparrow|)] \\
&= -i(|\uparrow\rangle\langle\downarrow| + |\downarrow\rangle\langle\uparrow|)(|\uparrow\rangle\langle\downarrow| - |\downarrow\rangle\langle\uparrow|) \\
&\quad + i(|\uparrow\rangle\langle\downarrow| - |\downarrow\rangle\langle\uparrow|)(|\uparrow\rangle\langle\downarrow| + |\downarrow\rangle\langle\uparrow|) \\
&= -i(-|\uparrow\rangle\langle\uparrow| + |\downarrow\rangle\langle\downarrow|) + i(|\uparrow\rangle\langle\uparrow| - |\downarrow\rangle\langle\downarrow|) \\
&= 2i(|\uparrow\rangle\langle\uparrow| - |\downarrow\rangle\langle\downarrow|) = 2i\sigma_z. \qquad \text{(B.13)}
\end{aligned}$$

The equalities $[\sigma_y, \sigma_z] = 2i\sigma_x$ and $[\sigma_z, \sigma_x] = 2i\sigma_y$ are shown similarly. Equation (3.20) is shown as

$$\begin{aligned}
\{\sigma_x, \sigma_y\} &= \sigma_x\sigma_y + \sigma_y\sigma_x \\
&= -i(-|\uparrow\rangle\langle\uparrow| + |\downarrow\rangle\langle\downarrow|) - i(|\uparrow\rangle\langle\uparrow| - |\downarrow\rangle\langle\downarrow|) \\
&= 0.
\end{aligned}$$

The equality $\{\sigma_y, \sigma_z\} = \{\sigma_z, \sigma_x\} = 0$ is shown in the same way.

[5] We have a commutation relation;

$$\begin{aligned}
[l_x, l_y] &= [yp_z - zp_y, zp_x - xp_z] \\
&= [yp_z, zp_x] + [zp_y, xp_z] \\
&= y[p_z, z]p_x + x[z, p_z]p_y \\
&= -iyp_x + ixp_y = l_z,
\end{aligned}$$

where we have made use of Eq. (3.2) and the relation $[ab, c] = a[b, c] + [a, c]b$. The equation $[l_y, l_z] = il_x$, $[l_z, l_x] = il_y$ can be derived in the same way.

[6] The polarized states $|\leftrightarrow\rangle$ and $|\updownarrow\rangle$ can be considered as three-dimensional vectors in the directions of the $x$- and $y$-axes, respectively. By rotating them around the $z$-axis by an angle $d\alpha$, we have

$$\begin{aligned}
D_z(d\alpha)|\leftrightarrow\rangle &= \cos d\alpha|\leftrightarrow\rangle + \sin d\alpha|\updownarrow\rangle \\
&\approx |\leftrightarrow\rangle + d\alpha|\updownarrow\rangle, \\
D_z(d\alpha)|\updownarrow\rangle &= \cos d\alpha|\updownarrow\rangle - \sin d\alpha|\leftrightarrow\rangle \\
&\approx |\updownarrow\rangle - d\alpha|\leftrightarrow\rangle,
\end{aligned}$$

where $d\alpha \ll 1$ has been assumed. Thus

$$\begin{aligned} D_z(d\alpha)|L\rangle &= \frac{1}{\sqrt{2}}\{|\updownarrow\rangle - d\alpha|\leftrightarrow\rangle - i|\leftrightarrow\rangle - id\alpha|\updownarrow\rangle\} \\ &= (1-id\alpha)\frac{1}{\sqrt{2}}\{|\updownarrow\rangle - i|\leftrightarrow\rangle\} \\ &= (1-id\alpha)|L\rangle, \end{aligned} \tag{B.14}$$

$$\begin{aligned} D_z(d\alpha)|R\rangle &= \frac{1}{\sqrt{2}}\{|\updownarrow\rangle - d\alpha|\leftrightarrow\rangle + i|\leftrightarrow\rangle + id\alpha|\updownarrow\rangle\} \\ &= (1+id\alpha)|R\rangle. \end{aligned} \tag{B.15}$$

By comparing Eq. (3.64): $D_z(d\alpha) \approx 1 - il_z d\alpha$ with Eqs. (B.14) and (B.15), we have

$$l_z|L\rangle = +1|L\rangle, \qquad l_z|R\rangle = -1|R\rangle.$$

Since a photon has spin of magnitude 1, the direction of propagation is parallel to the direction of spin in the circular polarized state $|L\rangle$, while the direction of propagation is anti-parallel to the direction of spin in the circular polarized state $|R\rangle$. Helicity $h$ is defined as

$$h = (\mathbf{l}\cdot\hat{\mathbf{p}}),$$

with the unit vector in the direction of momentum $\hat{\mathbf{p}}$. Since the circular polarized states $|L\rangle, |R\rangle$ are polarized in the $xy$-plane, the motion is in the direction of the $z$-axis:

$$\hat{\mathrm{p}} = \boldsymbol{\varepsilon}_1 \times \boldsymbol{\varepsilon}_2 = \boldsymbol{\varepsilon}_3.$$

Thus the helicity is expressed as

$$h = l_z,$$

and has values:

$$h(L) = +1, \qquad h(R) = -1.$$

## [Chapter 4]

**[1]** Equation (4.2) gives

$$\mathbf{S}^2 = \frac{1}{2}(3 + \boldsymbol{\sigma}_1\cdot\boldsymbol{\sigma}_2).$$

About $|\Psi^{(+)}\rangle$, the phase of the non-diagonal term is + in Eq. (4.4). Therefore, using

$$\begin{aligned} &\langle\Psi^{(+)}|\boldsymbol{\sigma}_1\cdot\boldsymbol{\sigma}_2|\Psi^{(+)}\rangle \\ &= {}_1\langle\uparrow|_2\langle\downarrow|\{\sigma_{x_1}\sigma_{x_2} + \sigma_{y_1}\sigma_{y_2}\}|\downarrow\rangle_1|\uparrow\rangle_2 + {}_1\langle\uparrow|_2\langle\downarrow|\sigma_{z_1}\sigma_{z_2}|\uparrow\rangle_1|\downarrow\rangle_2 \\ &= (1\cdot 1 - i\cdot i) + (1\cdot(-1)) = 1, \end{aligned}$$

we have

$$\langle \Psi^{(+)}|\mathbf{S}^2|\Psi^{(+)}\rangle = S(S+1) = 2.$$

This means that the magnitude of the total spin is $S = 1$.

**[2]** The three photon state can be expressed as

$$\begin{aligned}|\Psi\rangle_{123} &= |\phi\rangle_1|\Psi\rangle_{23}\\ &= (a|\uparrow\rangle_1 + b|\downarrow\rangle_1)\frac{1}{\sqrt{2}}\{|\uparrow\rangle_2|\downarrow\rangle_3 - |\downarrow\rangle_2|\uparrow\rangle_3\}\\ &= \frac{a}{\sqrt{2}}\{|\uparrow\rangle_1|\uparrow\rangle_2|\downarrow\rangle_3 - |\uparrow\rangle_1|\downarrow\rangle_2|\uparrow\rangle_3\}\\ &\quad + \frac{b}{\sqrt{2}}\{|\downarrow\rangle_1|\uparrow\rangle_2|\downarrow\rangle_3 - |\downarrow\rangle_1|\downarrow\rangle_2|\uparrow\rangle_3\} \qquad \text{(B.16)}\end{aligned}$$

By substituting

$$\begin{aligned}|\uparrow\rangle_1|\uparrow\rangle_2 &= \frac{1}{\sqrt{2}}\left\{|\Phi^{(+)}\rangle + |\Phi^{(-)}\rangle\right\},\\ |\downarrow\rangle_1|\downarrow\rangle_2 &= \frac{1}{\sqrt{2}}\left\{|\Phi^{(+)}\rangle - |\Phi^{(-)}\rangle\right\},\\ |\uparrow\rangle_1|\downarrow\rangle_2 &= \frac{1}{\sqrt{2}}\left\{|\Psi^{(+)}\rangle + |\Psi^{(-)}\rangle\right\},\\ |\downarrow\rangle_1|\uparrow\rangle_2 &= \frac{1}{\sqrt{2}}\left\{|\Psi^{(+)}\rangle - |\Psi^{(-)}\rangle\right\}\end{aligned}$$

in Eq. (B.16), we obtain Eq. (4.9).

**[3]** In the arrangement of **a**, **b**, **a′**, **b′** in Fig. 4.7, the correlation coefficients are given as

$$\begin{aligned}E_{ab} &= \frac{2(\phi-\theta)}{\pi} - 1,\\ E_{ab'} &= \frac{2(\phi'-\theta)}{\pi} - 1,\\ E_{a'b'} &= \frac{2(\phi'-\theta')}{\pi} - 1,\\ E_{a'b} &= \frac{2(\theta'-\phi)}{\pi} - 1.\end{aligned}$$

Therefore we have

$$\begin{aligned}S &= \left|\frac{2(\phi-\theta)}{\pi} - 1 - \left(\frac{2(\phi'-\theta)}{\pi} - 1\right)\right| + \left|\frac{2(\phi'-\theta')}{\pi} - 1 + \frac{2(\theta'-\phi)}{\pi} - 1\right|\\ &= 2\frac{\phi-\phi'}{\pi} + 2\left(1 - \frac{\phi-\phi'}{\pi}\right).\end{aligned}$$

The sign of the last line has been determined from the fact $\pi \geq \phi - \phi' \geq 0$ in the arrangement of the detectors. Thus we have

$$S = 2,$$

and the CHSH inequality is satisfied.

## [Chapter 5]

**[1]** From the left end of the tape, the value of the bit of $2^0$, the value of the bit of $2^1$, ..., are written on the cells as the input. In proceeding with computation, a value "1" is added to the left-end cell giving the least significant bit (cell). If the carry is not generated, the computation is finished. If the carry is generated, a value "1" is written on the control unit, and the head is moved to the right by one unit cell. The same procedure will be repeated continuously from the next cell until the computation is stopped.

- control unit: $q = 0, 1$
- tape mark : 0, 1
- operation rule: the following table.

| @state of control unit | tape mark | | |
|---|---|---|---|
| | 0 | 1 | B |
| 0 | Halt | Halt | Halt |
| 1 | 01R | 10R | 01R |

- initial state: $q = 1$ and the head placed at the left end.

## [Chapter 6]

**[1]** • About the Fredkin gate, Eq. (6.29) gives $A = A'$. Both $B$ and $C$ are unchanged if $A' = A = 0$, while $B$ and $C$ are interchanged if $A' = A = 1$. Thus the relation is the same as Eqs. (6.29)$\sim$(6.31), with the inputs and the outputs interchanged:

$$\begin{aligned} A &= A' \\ B &= (\overline{A'} \cdot B') + (A' \cdot C') \\ C &= (\overline{A'} \cdot C') + (A' \cdot B'). \end{aligned}$$

- About the Toffoli gate, it follows from Eqs. (6.33) and (6.34) that the inputs $A$ and $B$ are the same as the outputs $A'$ and $B'$, respectively. As for $C$, by applying $\oplus(A \cdot B)$ to Eq. (6.35), we have

$$\begin{aligned}\text{left-hand side} &= C' \oplus (A \cdot B) = C' \oplus (A' \cdot B') \\ \text{right-hand side} &= C \oplus (A \cdot B) \oplus (A \cdot B) \\ &= C \oplus ((A \cdot B) \oplus (A \cdot B)) \\ &= C \oplus 0.\end{aligned}$$

Thus we obtain $C' \oplus (A' \cdot B')$. The relation is again the same as Eqs. (6.33)→(6.35), with the inputs and the outputs interchanged:

**[2]** Figure B.1 shows an example.

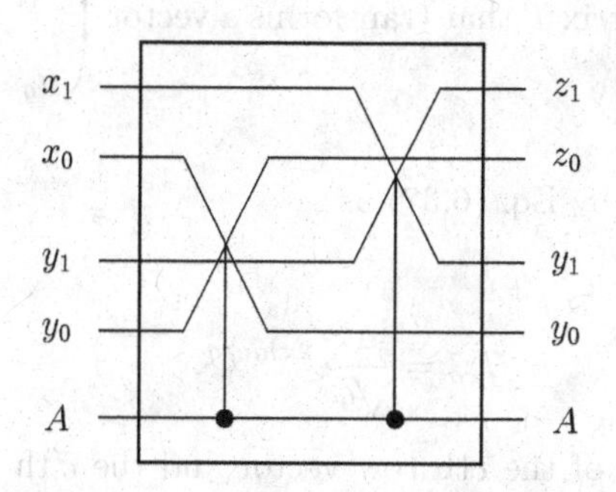


Fig. B.1 Multiplexer by the Fredkin gate.

**[3]** For example, referring to Example 6.4, one can design the circuit in Fig. B.2 that the output $A + B$ for given $A$ and $B$.

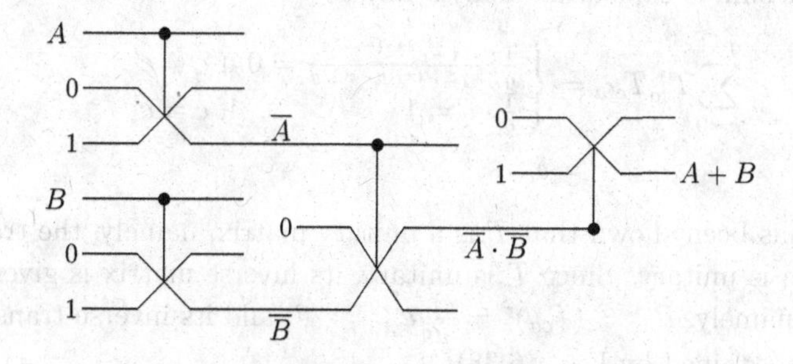


Fig. B.2 OR gate using the Fredkin gate.

**[4]** Figure B.3 shows an example.

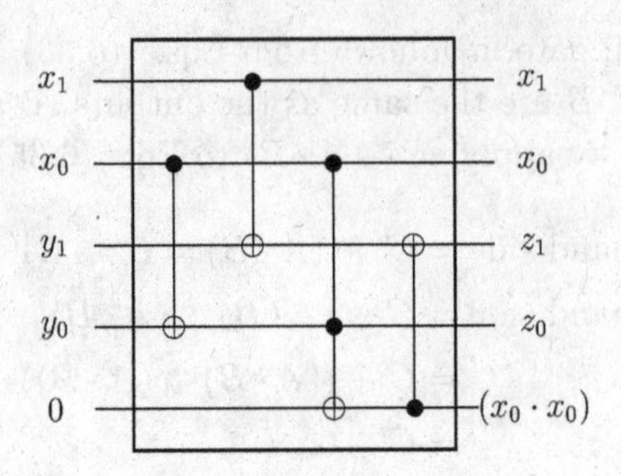


Fig. B.3 Adder for $Z = X + Y \pmod 4$.

**[5]** The elements of matrix $T$ that transforms a vector $\begin{pmatrix} u_0 \\ \vdots \\ u_{q-1} \end{pmatrix}$ to a vector $\begin{pmatrix} f_0 \\ \vdots \\ f_{q-1} \end{pmatrix}$ are given by Eq. (6.37) as

$$T_{ca} = \frac{1}{\sqrt{q}} e^{2\pi iac/q}.$$

The scalar product of the $c$th row vector and the $c'$th row vector becomes

$$\sum_{a=0}^{q-1} T^*_{ca} T_{c'a} = \frac{1}{q} \sum_{a=0}^{q-1} e^{2\pi ia(c-c')/q},$$

where the sum is calculated analytically as

$$\sum_{a=0}^{q-1} T^*_{ca} T_{c'a} = \begin{cases} \frac{1}{q} \cdot \frac{1-e^{2\pi i(c-c')}}{1-e^{2\pi i(c-c')/q}} = 0 & \text{if } c \neq c' \\ \frac{1}{q} \cdot q = 1 & \text{if } c = c' \end{cases}$$
$$= \delta_{cc'}.$$

Thus it has been shown that $T$ is a unitary matrix, namely, the transformation is unitary. Since $T$ is unitary, its inverse matrix is given as $R = T^\dagger$, namely, $R_{ac} = (T_{ca})^* = \frac{1}{\sqrt{q}} e^{-2\pi iac/q}$, and its inverse transformation is attained by Eq. (6.38).

**[6]** We consider the transformation on the basis:
$|a_2 a_1 a_0\rangle = |000\rangle, |001\rangle, |010\rangle, |011\rangle, |100\rangle, |101\rangle, |110\rangle, |111\rangle$.
The matrix corresponding to the first Hadamard transformation on $a_2$

component is given as

$$\frac{1}{\sqrt{2}}\begin{pmatrix} 1&0&0&0&1&0&0&0\\ 0&1&0&0&0&1&0&0\\ 0&0&1&0&0&0&1&0\\ 0&0&0&1&0&0&0&1\\ 1&0&0&0&-1&0&0&0\\ 0&1&0&0&0&-1&0&0\\ 0&0&1&0&0&0&-1&0\\ 0&0&0&1&0&0&0&-1 \end{pmatrix}. \tag{B.17}$$

The remaining parts are expressed by the following matrices:

$$\begin{pmatrix} 1&0&0&0&0&0&0&0\\ 0&1&0&0&0&0&0&0\\ 0&0&1&0&0&0&0&0\\ 0&0&0&1&0&0&0&0\\ 0&0&0&0&1&0&0&0\\ 0&0&0&0&0&1&0&0\\ 0&0&0&0&0&0&e^{i\pi/2}&0\\ 0&0&0&0&0&0&0&e^{i\pi/2} \end{pmatrix}, \quad \begin{pmatrix} 1&0&0&0&0&0&0&0\\ 0&1&0&0&0&0&0&0\\ 0&0&1&0&0&0&0&0\\ 0&0&0&1&0&0&0&0\\ 0&0&0&0&1&0&0&0\\ 0&0&0&0&0&e^{i\pi/4}&0&0\\ 0&0&0&0&0&0&1&0\\ 0&0&0&0&0&0&0&e^{i\pi/4} \end{pmatrix},$$

$$\frac{1}{\sqrt{2}}\begin{pmatrix} 1&0&1&0&0&0&0&0\\ 0&1&0&1&0&0&0&0\\ 1&0&-1&0&0&0&0&0\\ 0&1&0&-1&0&0&0&0\\ 0&0&0&0&1&0&1&0\\ 0&0&0&0&0&1&0&1\\ 0&0&0&0&1&0&-1&0\\ 0&0&0&0&0&1&0&-1 \end{pmatrix}, \quad \begin{pmatrix} 1&0&0&0&0&0&0&0\\ 0&1&0&0&0&0&0&0\\ 0&0&1&0&0&0&0&0\\ 0&0&0&e^{i\pi/2}&0&0&0&0\\ 0&0&0&0&1&0&0&0\\ 0&0&0&0&0&1&0&0\\ 0&0&0&0&0&0&1&0\\ 0&0&0&0&0&0&0&e^{i\pi/2} \end{pmatrix},$$

$$\frac{1}{\sqrt{2}}\begin{pmatrix} 1&1&0&0&0&0&0&0\\ 1&-1&0&0&0&0&0&0\\ 0&0&1&1&0&0&0&0\\ 0&0&1&-1&0&0&0&0\\ 0&0&0&0&1&1&0&0\\ 0&0&0&0&1&-1&0&0\\ 0&0&0&0&0&0&1&1\\ 0&0&0&0&0&0&1&-1 \end{pmatrix}, \quad \begin{pmatrix} 1&0&0&0&0&0&0&0\\ 0&0&0&0&1&0&0&0\\ 0&0&1&0&0&0&0&0\\ 0&0&0&0&0&0&1&0\\ 0&1&0&0&0&0&0&0\\ 0&0&0&0&0&1&0&0\\ 0&0&0&1&0&0&0&0\\ 0&0&0&0&0&0&0&1 \end{pmatrix}.$$

Starting, for example, with Eq. (B.17), by multiplying those matrix to the left side successively, one obtains the matrix with its matrix elements $\frac{1}{\sqrt{8}}e^{2\pi iac}$ ($a = 0, 1, ..., 7$, $c = 0, 1, ..., 7$).

## [Chapter 7]

[1] About the maximum, noting $\sum_{i=1}^{N} p_i = 1$, let us show the inequality

$$I \leq \log_2 N, \tag{B.18}$$

making use of the relation

$$\ln x \leq x - 1. \tag{B.19}$$

We have first the relation

$$\begin{aligned}
\left(-\sum_{i=1}^{N} p_i \ln p_i\right) - \ln N &= \left(-\sum_{i=1}^{N} p_i \ln p_i\right) - \left(\sum_{i=1}^{N} p_i\right) \ln N \\
&= -\sum_{i}^{N} p_i(\ln p_i + \ln N) \\
&= \sum_{i=1}^{N} p_i \ln\left(\frac{1}{Np_i}\right) \\
&\leq \sum_{i=1}^{N} p_i \left(\frac{1}{Np_i} - 1\right) \\
&= \frac{1}{N}\left(\sum_{i=1}^{N} 1\right) - \sum_{i=1}^{N} p_i \\
&= 0.
\end{aligned} \tag{B.20}$$

The inequality (B.18) is obtained by changing the base of logarithm from $e$ to 2. The equality holds when $p_1 = p_2 = \ldots = p_N = 1/N$. Therefore, the maximum is $\log_2 N$. The minimum value 0 is obtained if one value has $p_j = 1$ and all other values $i \neq j$ are zero in the sum $\sum_{i=1}^{N} p_i = 1$.

## [Chapter 8]

[1] Using the expansion: $a = a_{n-1}2^{n-1} + a_{n-2}2^{n-2} + \cdots + a_1 2^1 + a_0 2^0$, one writes $x^a$ as

$$x^a = x^{a_{n-1}2^{n-1}} x^{a_{n-2}2^{n-2}} \cdots x^{a_1 2^1} x^{a_0 2^0}. \tag{B.21}$$

Note that with two pairs of integers: $a$, $a'$, $b$, $b'$, satisfying

$$\begin{aligned}
a &\equiv a' \bmod N \\
b &\equiv b' \bmod N.
\end{aligned}$$

The multiplication theorem with congruence equations (A.18) states that

$$ab \equiv a'b' \bmod N. \tag{B.22}$$

By applying this multiplication theorem to Eq. (B.21), we have

$$\begin{aligned} x^{a_{n-1}2^{n-1}} &\equiv x'_{n-1} \bmod N \\ x^{a_{n-2}2^{n-2}} &\equiv x'_{n-2} \bmod N \\ &\vdots \\ &\vdots \\ x^{a_1 2^1} &\equiv x'_1 \bmod N \\ x^{a_0 2^0} &\equiv x'_0 \bmod N. \end{aligned}$$

Then, we have the relation:

$$\begin{aligned} x^a &= x^{a_{n-1}2^{n-1}} x^{a_{n-2}2^{n-2}} \cdots x^{a_1 2^1} x^{a_0 2^0} \\ &\equiv x'_{n-1} x'_{n-2} \cdots x'_1 x'_0 \bmod N. \end{aligned}$$

In this way, Equation (8.83) is proved.

[2] Set the initial state as

$$|\psi_1\rangle = |0\rangle|0\rangle|0\rangle, \tag{B.23}$$

where both the first and the second registers have $n$ bits.

**Step 1**:

Apply the Hadamard transformation to the first and the second registers. The result is

$$|\psi_2\rangle = \frac{1}{q} \sum_{x_1,x_2=0}^{q-1} |x_1\rangle|x_2\rangle|0\rangle, \tag{B.24}$$

where $q = 2^n$.

**Step 2**:

Apply the Oracle operator $U$ to $|\psi_2\rangle$:

$$|\psi_3\rangle = U|\psi_2\rangle = \frac{1}{q} \sum_{x_1,x_2=0}^{q-1} |x_1\rangle|x_2\rangle|f(x_1,x_2)\rangle. \tag{B.25}$$

Now, let $s$ be the solution of Eq. (8.84). Because $f(x_1, x_2) = b^{x_1} a^{x_2} = a^{sx_1+x_2} \bmod N$, we have

$$|\psi_3\rangle = \frac{1}{q} \sum_{x_1,x_2=0}^{q-1} |x_1\rangle|x_2\rangle|a^{sx_1+x_2} \bmod N\rangle. \tag{B.26}$$

Using Eq. (8.48)

$$|a^{sx_1+x_2} \bmod N\rangle = \sum_{t=0}^{r-1} \frac{1}{\sqrt{r}} e^{2\pi i(sx_1+x_2)t/r}|\Phi_t\rangle, \tag{B.27}$$

one rewrites the third register and obtains

$$|\psi_3\rangle = \frac{1}{q}\frac{1}{\sqrt{r}}\sum_{t=0}^{r-1}\left[\sum_{x_1=0}^{q-1} e^{2\pi i s x_1 t/r}|x_1\rangle \sum_{x_2=0}^{q-1} e^{2\pi i x_2 t/r}|x_2\rangle|\Phi_t\rangle\right]. \tag{B.28}$$

Thus, Fourier transformations of the states $|x_1\rangle$ and $|x_2\rangle$ have been accomplished, where $r$ is the order that satisfies $a^r \equiv 1 \bmod N$.

**Step 3**:

Perform Fourier transforms of $x_1$ and $x_2$;

$$\begin{aligned}|\psi_4\rangle &= \mathrm{QFT}^\dagger_{x_1}\mathrm{QFT}^\dagger_{x_2}|\psi_3\rangle \\ &= \frac{1}{q^2}\frac{1}{\sqrt{r}}\sum_{t=0}^{r-1}\left[\sum_{x_1,y_1=0}^{q-1} e^{2\pi i x_1(st/r-y_1/q)}|y_1\rangle\right] \\ &\quad\times\left[\sum_{x_2,y_2=0}^{q-1} e^{2\pi i x_2(t/r-y_2/q)}|y_2\rangle\right]|\Psi_t\rangle. \end{aligned} \tag{B.29}$$

By taking the summations of $x_1$ and $x_2$ in Eq. (B.29), we have

$$|\psi_4\rangle = \frac{1}{\sqrt{r}}\sum_{t=0}^{r-1}\left|\widetilde{st\frac{q}{r}}\right\rangle\left|\widetilde{t\frac{q}{r}}\right\rangle|\Psi_t\rangle. \tag{B.30}$$

Thus, one observes different periods in the first and the second registers. From the ratio of these periods, the solution $s$ of a discrete logarithm problem is determined.

## [Chapter 9]

[1] Using

$$\cos\left(\frac{\pi}{2}-\phi_A\right) = \sin\phi_A, \qquad \cos\phi_A + \sin\phi_A = \sqrt{2}\cos\left(\phi_A - \frac{\pi}{4}\right),$$

we can rewrite two terms of the left-hand side as

$$\begin{aligned}&\cos(-\phi_A)\cos\left(\frac{\pi}{4}-\theta_B\right) + \cos\left(\frac{\pi}{2}-\phi_A\right)\cos\left(\frac{\pi}{4}-\theta_B\right) \\ &= \sqrt{2}\cos\left(\phi_A-\frac{\pi}{4}\right)\cos\left(\frac{\pi}{4}-\theta_B\right).\end{aligned}$$

For the remaining terms on the left-hand side, from $\sin\phi_A - \cos\phi_A = \sqrt{2}\sin\left(\phi_A - \frac{\pi}{4}\right)$, we have

$$
\begin{aligned}
&-\cos(-\phi_A)\cos\left(\frac{3}{4}\pi - \theta_B\right) + \cos\left(\frac{\pi}{2} - \phi_A\right)\cos\left(\frac{3}{4}\pi - \theta_B\right) \\
&= \sqrt{2}\sin\left(\phi_A - \frac{\pi}{4}\right)\cos\left(\frac{3}{4}\pi - \theta_B\right) \\
&= \sqrt{2}\sin\left(\phi_A - \frac{\pi}{4}\right)\sin\left(\theta_B - \frac{\pi}{4}\right) \\
&= -\sqrt{2}\sin\left(\phi_A - \frac{\pi}{4}\right)\sin\left(\frac{\pi}{4} - \theta_B\right).
\end{aligned}
$$

Eventually, the right-hand side of Eq. (9.70) is given as

$$
\begin{aligned}
&\sqrt{2}\cos\left(\phi_A - \frac{\pi}{4}\right)\cos\left(\frac{\pi}{4} - \theta_B\right) - \sqrt{2}\sin\left(\phi_A - \frac{\pi}{4}\right)\sin\left(\frac{\pi}{4} - \theta_B\right) \\
&= \sqrt{2}\cos(\phi_A - \phi_B).
\end{aligned}
$$

**[2]**

(1) From Fig. 9.6, we have

$$
\begin{aligned}
|P(\mathbf{a},\mathbf{b}) - P(\mathbf{a},\mathbf{b}')| &= |\cos(\alpha+\beta) - \cos(\alpha-\beta)| \\
&= 2|\sin\alpha\sin\beta|.
\end{aligned}
$$

Therefore

$$
\begin{cases}
\text{maximum: } 2\sin\beta & \text{with } \alpha = \pm\pi/2 \\
\text{minimum: } 0 & \text{with } \alpha = 0,\ \pi
\end{cases}
$$

(2) From Fig. 9.6, we have

$$
\begin{aligned}
|P(\mathbf{a}',\mathbf{b}) + P(\mathbf{a}',\mathbf{b}')| &= |\cos(\alpha'+\beta) + \cos(\alpha'-\beta)| \\
&= 2|\cos\alpha'\cos\beta|.
\end{aligned}
$$

Therefore

$$
\begin{cases}
\text{maximum: } 2\cos\beta & \text{with } \alpha' = 0,\ \pi \\
\text{minimum: } 0 & \text{with } \alpha' = \pm\pi/2
\end{cases}
$$

(3) From the above arguments, we conclude

- The maximum is obtained by setting $\alpha = \pm\pi/2$ and $\alpha' = 0$, or $\pi$:

$$2\sin\beta + 2\cos\beta = 2\sqrt{2}\sin(\beta + \pi/4).$$

  Therefore $\beta = \pi/4$ gives the maximum $2\sqrt{2}$.
- The minimum is attained by setting $\alpha = 0$, or $\pi$, and $\alpha' = \pm\pi/2$, where the minimum is 0 independent of the value of $\beta$.

## [Chapter 10]

[1] From $\sin\frac{\theta}{2} = \frac{1}{\sqrt{8}}, \cos\frac{\theta}{2} = \sqrt{\frac{7}{8}}$, using the help of the trigonometric addition formulas successively, we have

$$\sin\theta = 2\sin\frac{\theta}{2}\cos\frac{\theta}{2} = \frac{\sqrt{7}}{4}, \quad \cos\theta = \frac{3}{4},$$

$$\sin\left(\frac{3}{2}\theta\right) = \sin\frac{\theta}{2}\cos\theta + \cos\frac{\theta}{2}\sin\theta = \frac{5}{4\sqrt{2}}, \quad \cos\frac{3}{2}\theta = \frac{1}{4}\sqrt{\frac{7}{2}},$$

$$\sin\left(\frac{5}{2}\theta\right) = \sin\frac{3\theta}{2}\cos\theta + \cos\frac{3\theta}{2}\sin\theta = \frac{11}{8\sqrt{2}}.$$

As a result, we have

$$P(z_0)_{k=1} = \frac{25}{32}$$

$$P(z_0)_{k=2} = \frac{121}{128}.$$

## [Chapter 11]

[1] By Taylor-expansion of a function $f(|\mathbf{R}-\mathbf{r}|)$ around $\mathbf{R}$ with deviation $-\mathbf{r}$, we have

$$f(|\mathbf{R}-\mathbf{r}|) \approx f(|\mathbf{R}|) + \sum_i \frac{\partial f(|\mathbf{R}|)}{\partial X_i}(-x_i). \tag{B.31}$$

Making use of $f(|R|) = 1/|R|$, we have

$$\frac{\partial f(|\mathbf{R}|)}{\partial X_i} = \frac{\partial}{\partial X_i}\frac{1}{|\mathbf{R}-\mathbf{r}|}\bigg|_{r=0} = -\frac{X_i}{|\mathbf{R}-\mathbf{r}|^3}\bigg|_{r=0} = -\frac{X_i}{R^3}. \tag{B.32}$$

Then, it follows that

$$\frac{1}{|\mathbf{R}-\mathbf{r}|} \approx \frac{1}{R} + \sum_i \frac{X_i x_i}{R^3} = \frac{1}{R} + \frac{(\mathbf{R}\cdot\mathbf{r})}{R^3}. \tag{B.33}$$

# Bibliography

We list the textbooks and other references which we looked up in writing this book. There are quite many publications recently about quantum information. In the following, we will give only those books that especially enlightened us, and that we strongly recommend to read for further studies. As a textbook covering the whole area of quantum information,


- M. Nielsen and I. Chuang, "Quantum Computation and Quantum Information" (Cambridge University Press, 2000)


has the reputation of being a comprehensive book.


- Colin P. Williams and Scott H. Clearwater "Explorations in Quantum Computing" (Springer, New York, 1997)


presents many examples studied with Mathematica.


- Dirk Bouwmeester, Artur Ekert and Anton Zeilinger (eds.), "The Physics of Quantum Information" (Springer, Berlin, 2000)


gives detailed explanations of the experiments in quantum computers. As for Shor's algorithm, the review article


- A. Ekert and R. Jozsa, Reviews of Modern Physics 68, pp. 733-753 (1996)


is carefully written and intelligible.

For specific subjects, we recommend the readers to try the original papers. For those readers, we will refer to the website of the preprint library of Los Alamos: `http://xxx.lanl.gov` where papers in variety of fields are available from those archived according to year.

There are many textbooks in quantum mechanics. Those describing the EPR pair and the theory of measurement are, however, limited, among which, the books:

- Daniel R. Bes "Quantum Mechanics" (Springer, 2007)
- J. J. Sakurai, "Modern Quantum Mechanics," Revised ed. (Addison-Wesley, Reading, 1994)
- J. J. Sakurai, "Advanced Quantum Mechanics" (Addison-Wesley, Reading, 1967)
- A. Peres, "Quantum theory: Concepts and Methods" (Kluwer, Dordrecht, 1993)

have been helpful. In addition to the two books by J. J. Sakurai given above, we mention another general textbook in quantum mechanics,

- A. Messiah, "Quantum Mechanics" (Dover, N. Chemsford, 1999)

which is excellent for practical purposes. As a book describing interesting aspects in the history of cryptography,

- Simon Singh, "The Code Book" (Fourth Estate, New York, 1999)

is worth reading.

# Index